Advances in Anatomy
Embryology and Cell Biology

Vol. 77

W0050439

Editors
F. Beck, Leicester W. Hild, Galveston
J. van Limborgh, Amsterdam R. Ortmann, Köln
J.E. Pauly, Little Rock T.H. Schiebler, Würzburg

Advances in Anatomy
Embryology and Cell Biology

Vol. 72

Eva Braak

On the Structure of the
Human Striate Area

With 44 Figures

Springer-Verlag
Berlin Heidelberg New York 1982

Dr. Eva Braak, PD
Dr. Senckenbergisches Anatomisches Institut
Zentrum der Morphologie
Johann-Wolfgang-Goethe-Universität
Theodor-Stern-Kai 7
D-6000 Frankfurt/M. 70
FRG

This work was done in the Anatomischen Institut of the Christian-Albrecht-Universität in Kiel (Head: Prof. Dr. med. H. Leonhardt). It is a revised version of the author's *Habilitationsschrift* submitted to the Medical Faculty of the Christian-Albrecht-Universität of Kiel 1978. Title of the German *Habilitationsschrift: Licht- und elektronenmikroskopische Untersuchungen zur Morphologie der primären Sehrinde des Menschen.*

ISBN-13:978-3-540-11512-0   e-ISBN-13:978-3-642-68572-9
DOI: 10.1007/978-3-642-68572-9

Composition: Schreibsatz-Service Weihrauch, Würzburg

2121/3321-543210

Contents

1 Introduction

Primary cortical areas receive a defined input which makes them especially appropriate for investigating cortical functions. The striate area is the only isocortical field which can be delineated unequivocally in the human brain. Nevertheless, there have been only a few morphological studies of this particular area (cytoarchitectonic studies: Bailey and Von Bonin 1951, Beck 1934, Von Economo and Koskinas 1925, Filimonoff 1932; myeloarchitectonic studies: Sanides and Vitzthum 1965, Vogt and Vogt 1919; pigmentoarchitectonic studies: Braak H 1976, 1977). For Golgi impregnations, Ramón y Cajal (1900, 1909–1911), Conel (1939–1967), and Shkol'nik-Yarros (1971) preferred the incompletely myelinated material taken from brains of young children – a fact that somewhat restricts their descriptions of the human striate area.

Pigment preparations (Braak H 1978) provide a detailed view of the lamination of cortical areas. Furthermore, many types of cortical nerve cells reveal a typical lipofuscin-pigment pattern (Braak H 1974a). Thus, a correlation can be drawn between the type of neuron as classified in Golgi preparations and the characteristic number and distribution of lipofuscin granules found in the cell body. Neurolipofuscin granules can therefore be considered the internal markers. In this study several cell types of the striate area have been identified under light and electron microscopes by means of their characteristic pigmentation.

Until now there has been no electron-microscopical study which describes all the cell types occurring in the various laminae of the human cerebral cortex. The few descriptions available are confined to parts of the cortical band (Cragg 1976; Braak E 1975, 1976a, 1980; Lopes and Mair 1974, Ramsey 1965).

On the other hand, there are numerous studies dealing with the cortical morphology of subhuman primates and other experimental animals. The primary visual area in particular has been carefully studied in a great number of mammalian species. It is apparent from these studies that there are considerable differences between the human striate area and that of lower mammals.

The aim of this investigation is to present a detailed description of the cellular constituents of the adult human striate area through light and electron microscopy. Differences and conformities with the former studies concerning the immature human striate area and the primary visual field of subhuman primates will be referred to in particular.

2 Material and Methods

Human cortical tissue from Area 17 was obtained by surgery from two women (32-year-old: glioblastoma multiforme; 66-year-old: metastatic tumor) and three men (36-year-old: glioma; 66-year-old: malignant tumor; 72-year-old: glioblastoma multiforme).

Immediately after removal, the samples were put into a freshly prepared mixture of glutaraldehyde and formaldehyde in phosphate buffer at pH 7.2 (Sotelo 1969). After 5–20 min they were cut with a razor blade into slices approximately 2–3 mm in thickness for light-microscopical investigations and 1–1.5 mm in thickness for electron-microscopical studies.

For electron microscopy the slices were transferred to fresh fixative fluid for about 6–8 h at 4 °C, and thereafter cut, perpendicularly to the cortical surface, into pieces less than 1 mm wide, comprising the entire depth of the cortical gray and a small part of the white matter. The pieces were rinsed in phosphate buffer for about 10 h at pH 7.2 at 4 °C, the buffer solution being changed three times. Pieces were taken from the surface of the gyrus, from the border between gyrus and sulcus, and from the wall and the groove of the sulcus. They were postfixed in 2% OsO₄ for 2–4 h at 4 °C. Some of them were stained en bloc with uranyl acetate (Karnovsky 1967), and embedded in Araldite.

For the most desirable radial orientation, 2-μm-thick sections were cut with an LKB pyramitome or with a Reichert OmU2 microtome and stained with methylene blue-azure II (Richardson et al. 1960). The blocks were trimmed to include either the upper half or the lower half of the cortex. The ultrathin sections were grid stained with saturated uranyl acetate in 70% methanol, followed by lead citrate (Reynolds 1963).

For light-microscopical investigations, the 2–3 mm thick brain slices were stored for 8 days at 4 °C in the fixative fluid. For the Nissl preparations, embedding was carried out as for electron microscopy, but without osmication and uranyl acetate staining. Sections of 10 μm thickness were cut with a Jung Autocut and stained for half an hour with methylene blue (Braak E 1976b). Golgi impregnations were done according to the modification of Braitenberg et al. (1967). The sections cut on a freezing microtome at 100 μm were rapidly dehydrated, cleared in xylene and mounted with synthetic resin (Caedax, Merck or Permount, Fisher). In Golgi impregnations the layers are defined by counterstaining several sections with neutral red according to a method described by H. Braak (1974a). A Zeiss and a Leitz photomicroscope (Agfa ortho 25 professional emulsion) and a Siemens Elmiskop 101 (operated at 80 kV; Agfa Scientia 23 D 56 and DuPont Graphic arts film Cronar ortho S litho) were used.

Throughout the description and the microphotographs the pial surface is referred to as the uppermost, the white matter as the lowermost boundary of the cortex; hence higher or ascending means toward the pia; below, lower, or descending means toward the white matter; and horizontal means staying within a defined layer.

For determining the size classes of the neurons, I basically used the classification of Von Economo and Koskinas (1925) for pyramidal cells – small: cell body height 10–15 μm, width 7–10 μm, nucleus diameter 7–8 μm, medium sized: cell body height 20–30 μm, width 8–10 μm, nucleus diameter 8–10 μm; large: cell body height 30–50 μm, width 15–20 μm, nucleus diameter about 12 μm.

3 Lamination Pattern

Berlin (1858), a pupil of Meynert (1867, 1868), was the first to describe the six layers within the cerebral isocortex. The layers are distinguished from each other by the relative number, size, and types of their neurons. Glial and endothelial cells do not contribute significantly to the gray level indices of the layers (Fleischhauer and Vossel 1979, rabbit; Wree et al. 1980, *Callithrix jacchus*). On the recommendation of Vogt and Vogt (1919) the lamination is designated by Roman numerals according to cytoarchitectonic methods, as seen in Nissl preparations. Considering the differences in terminology used by various authors in the laminar differentiation of the human striate area, it seems advisable to give a juxtaposition of the more commonly encountered schemes. For the comparison of the various lamination patterns, the location of the stria of Gennari is a landmark. On the assumption that the stria of Gennari has the same extent and localization in primates as in humans, the lamination patterns of some experimental monkeys are included in Table 1. The classification in particular of the subdivisions of the lower tier of the third layer and of those of the fourth layer is not uniform (Bil-

Table 1. Classification of the lamination in the striate area of man and various primates according to various authors. The grey area gives the approximate location of the stria of Gennari

Ramón y Cajal (1900) man	Brodmann (1903) man	Von Bonin (1942) man	Hassler and Wagner (1965) *Saimiri*	Hubel and Wiesel (1972) macaque	Lund (1973) macaque	Braak (1976) man
1	I	I	I	I	I	I
2	II	II	II	II	II	II
3	III	III A	III a	III	III	III ab
	IV a	III B	III b	IV a	IV a	IIIc/IVa
4	IV b	IV a	III c	IV b	IV b	IV b
					IV cα	IV cα
5	IV c	IV b	IV	IV c	IV cβ	IV cβ / IVd/Va
6	V	V a	V	V	V	V b
7		V b				
8	VI a	VI	VI	VI	VI	VI a
9	VI b					VI b

lings-Gagliardi et al. (1974), Von Economo and Koskinas (1925), Garey (1971), Valverde (1977). For the present investigation the lamination pattern proposed by H. Braak (1976) was adopted, which is based additionally on pigmentoarchitectonic studies.

4 Neurons and Neuropil of Layer I

4.1 Nissl-Stained and Methylene Blue-Azure II-Stained Sections

Layer I of the isocortex is conspicuous because of the very small number of neuronal perikarya. With the membrana limitans gliae superficialis, layer I touches the subarachnoid space. The lower border is defined by the abrupt appearance of numerous, densely packed, small cell bodies of the second layer (Figs. 1, 2, 10).

A distinct boundary can be drawn between the membrana limitans gliae superficialis and the subjacent neuropil of layer I (Fig. 2, Braak E 1975). Occasionally the membrana limitans glia superficialis is only faintly tinged. In some specimens vacuolation occurs within the most superficial neuropil; this is probably an ischemic alteration (Garcia et al. 1977, 1978; Jenkins et al. 1979; Kalimo et al. 1977).

The subjacent neuropil, the sublamina tangentialis, contains abundant cross- and longitudinally sectioned myelinated axons (Fig. 2, Fleischhauer and Laube 1977).

3

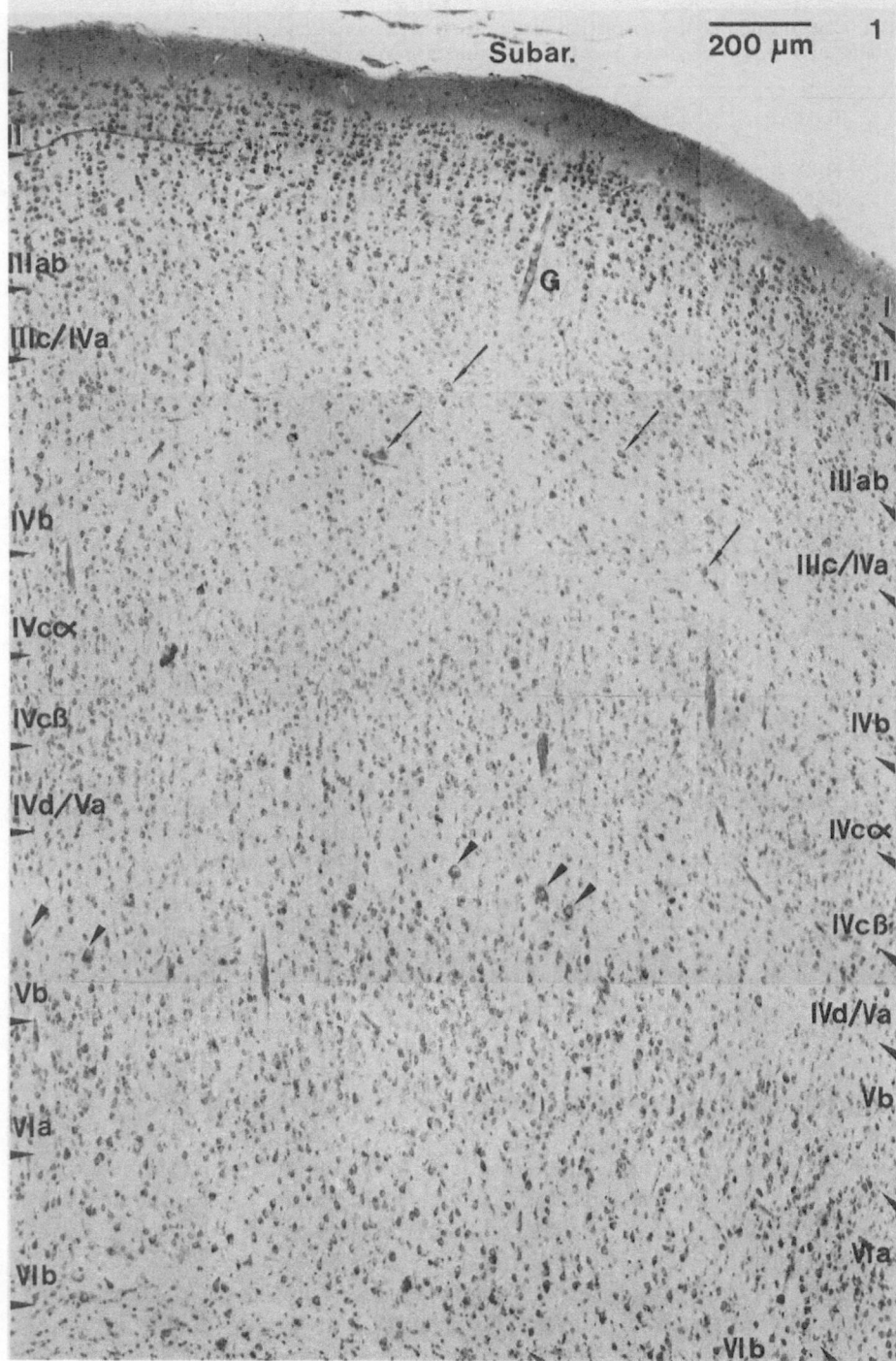

Fig. 1. Nissl-stained section through the human striate area (Araldite, 10 μm, methylene blue; 66-year-old man), using the lamination scheme of H. Braak (1976). *Subar.*, subarachnoid space. The cell-poor layer I is followed by the cell-rich layer II. This layer is hardly separable from the adjacent layer IIIab, in which its lower tier has a reduced number of cell bodies. In layer IIIc/IVa

Here and within the following sublamina infratangentialis a few cell bodies are scattered. Most of them are fibrillary astrocytes, a few can be considered light oligodendrocytes, and some are classified as type-II cells (Braak E 1975), which are presumably microglial cells (for details see Sect. 12). Nerve cells are rarely found.

The nerve cells (7–10 × 8–12 μm) exhibit ovoid nuclei, which have a smooth contour and contain chromatin condensations scattered lightly throughout (Figs. 3–5). The nuclei are surrounded by an intensely stained, narrow cytoplasmic rim which gives rise to one or two processes (Fig. 3). Satellite glial cells may be associated (Figs. 4, 5), but it is often impossible to classify them.

4.2 Golgi Preparations

The nerve cells of layer I, impregnated according to the Golgi technique, have a more or less ovoid cell body from which a few processes emerge (Fig. 6). Often the impregnation of the processes ceases near the cell body. There is only one example in which the processes can be traced for about 70–100 μm, showing irregularly spaced dilations and constrictions, as well as a few short side branches and bulbous endings. The processes resemble one another to such an extent as not to permit distinction between axons and dendrites.

The neurons of the first isocortical layer are more often successfully impregnated in fetal and early postnatal stages than in the adult brain. There are large neurons, usually referred to as the fetal Cajal-Retzius horizontal cells, and smaller ones, bipolar and polygonal neurons (Baron and Gallego 1971, rabbit; Bradford et al. 1977, rat; Fox et al. 1966, dog; Marin-Padilla 1972, cat; Noback and Purpura 1961, rat, cat, rabbit; Ramón y Cajal 1890, 1891, 1893, 1909–1911, 1929, dog, rabbit, cat, rat, mouse, man; Retzius 1891, 1893, 1894b, man, dog; Sas and Sanides 1970, dog, cat, hedgehog). The large fetal Cajal-Retzius horizontal cells have never been seen impregnated in adult human brains. It is uncertain whether the fetal Cajal-Retzius cells persist or not. There are two possibilities concerning the origin of layer-I neurons. The first is that the fetal Cajal-Retzius horizontal cells undergo a reduction in size and number of processes. The second is that all layer-I neurons may be derived from the smaller bipolar and polygonal neurons. To the best of our knowledge, Golgi-stained layer-I neurons have hitherto not been demonstrated in the adult human isocortex. The few impregnated neurons of layer I in adult experimental animals (Fox and Inman 1966, Sas and Sanides 1970, Sousa-Pinto et al. 1975) show dendritic spines and varicosities, whereas in man, the processes are spineless. Among the few neurons seen, a classification is not yet possible.

---

the number of cell bodies increases; they are of various sizes. This is in contrast with the next layer, IVb, which has few cell bodies. This zone contains some large Meynert-Cajal solitary neurons (*arrows*) and the stria of Gennari, which is not visible with this method. Layers IVcα and IVcβ are both cell rich. Layer IVcβ seems to be populated by almost one cell type only: small and faintly tinged. In layer IVd/Va the cell bodies tend to be arranged in columns and their number is slightly reduced. Layer Vb is characterized by the large Meynert pyramidal cells (*arrowheads*) and a low density of cell bodies. Layers VIa and VIb again are cell-rich zones. *G* blood vessel

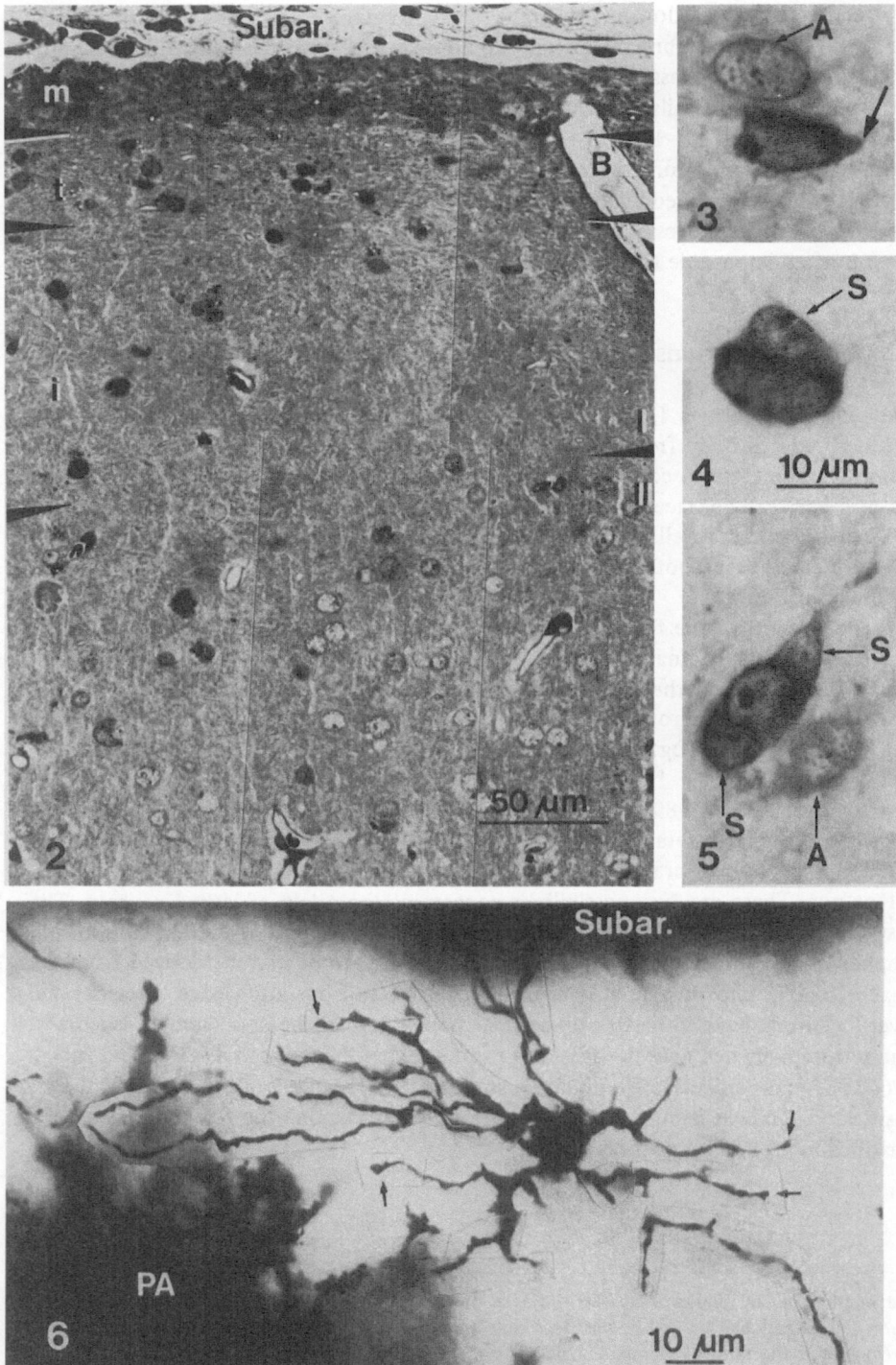

Fig. 2. Semithin section through the human striate area showing layers I and II (Araldite, 2 μm, methylene blue-azure II; 36-year-old man). Layer I comprises the membrana limitans gliae superficialis (*m*), the sublamina tangentialis (*t*), which contains numerous myelinated fibers, and the sub-

4.3 Electron Microscopy

Neurons

The nerve cell body has a smooth cell contour, from which a process rarely emerges. The nuclei are deeply indented; they contain a few small chromatin condensations (Fig. 7). The perinuclear cistern is frequently in continuity with the cisterns of the rough endoplasmic reticulum (RER; Fig. 8). Mitochondria, Golgi complexes, microtubules and lipofuscin granules with small lipid droplets are poorly developed. Polyribosomes are dispersed among these organelles. Occasionally a cilium is visible. In a few cases, medium-sized boutons with pleomorphic vesicles make axosomatic synaptic contacts, of which the postsynaptic thickening is absent or thin. Often a satellite glial cell is associated (Fig. 8); if it is a light oligodendrocyte, the facing membranes may be connected by puncta adhaerentia.

The neurons of the marginal zone, the subsequent layer I, are present before the cortical plate is visible (preplate neuron, König et al. 1977; Rickmann et al. 1977, rat). This raises the question of whether the layer-I neurons of the adult cortex can be classified in terms of the two main groups of cortical neurons: the pyramidal and nonpyramidal, or stellate, cells, which both develop from the cortical plate. Sloper (1973, rhesus monkey, motor cortex) characterized small nonpyramidal cells and found that they occurred most commonly in layer II, but also in all the other layers, including layer I. Electronmicroscopically, layer I neurons in the adult human striate area have features mildly reminiscent of small nonpyramidal cells (see Sect. 11.3.), i.e., the indented nucleus and the continuity of the perinuclear cistern with the cisterns of the rough endoplasmic reticulum, as well as the very rare axosomatic synapses. Hence, I tend to classify these cells within the group of nonpyramidal cortical nerve cells.

Neuropil

The external glial layer is a feltwork composed of different types of processes of fibrillary astrocytes (Braak E 1975). There is a fairly sharp transition from this layer to the subjacent neuropil of the sublamina tangentialis: the number of glial processes is reduced, and myelinated axons (diameter $0.5-5.0$ μm) and profiles of neuronal processes appear in abundance. Some conspicuously large boutons (diameter 2 μm) make asymmetric (Colonnier 1968) synaptic contacts on dendritic spines (diameter

---

lamina infratangentialis (*i*). Layer II starts abruptly with a great number of cell nuclei. *Subar.*, subarachnoid space; *B*, blood vessel

Figs. 3–5. Examples of Nissl-stained layer-I neurons (Araldite, 10 μm, methylene blue; 66-year-old man). In Figs. 3 and 5 a large astrocyte (*A*) is located near the neuron. The nerve cells of Figs. 4 and 5 are associated with one and two satellite cells (*S*) respectively. Furthermore, in Fig. 3 the proximal part of a process is visible (*arrow*)

Fig. 6. Photomicrograph of a Golgi-impregnated layer-I nerve cell (frozen section, 100 μm, Permount; 72-year-old man). The perikaryon gives rise to several processes, some of which have bulbous endings (*arrows*). *Subar.*, subarachnoid space; *PA*, protoplasmic astrocyte

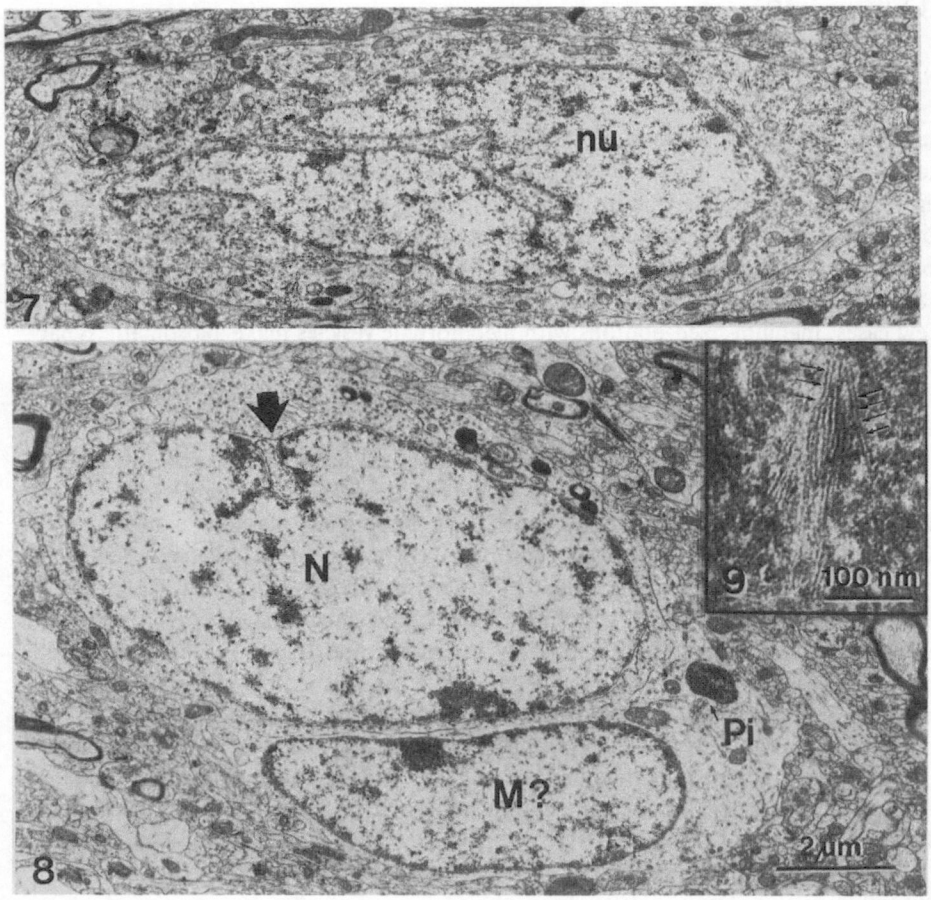

Fig. 7. Electron micrograph of a layer-I nerve cell with a deeply invaginated nucleus (*nu*) (66-year-old woman). For the scale see Fig. 8

Fig. 8. Layer-I neuron with the perinuclear cistern extending into cisterns of the RER (*arrow*). The cell body is associated with a satellite glial cell, which is tentatively identified as a microglial cell (*M?*). *Pi*, dense body. (66-year-old woman)

Fig. 9. The high magnification of the dense body of Fig. 8 reveals tubular subunits (*arrows*)

about 1 μm) with a prominent spine apparatus (Gray 1959). The boutons contain small spherical vesicles (diameter about 50 nm) and the active zone is interrupted. Within the lower portion of layer I (sublamina infratangentialis) the number of mye-linated axons is considerably reduced. Numerous smaller boutons containing spheri-cal vesicles make synaptic contacts with various dendritic profiles. Boutons with flat, pleomorphic, or dense-core vesicles rarely occur. A further characteristic feature of layer I is clusters of abundant, tiny, round profiles (diameter about 0.1 μm) which contain only one or two microtubules.

Most of the dendritic profiles belong to the end ramifications of pyramidal cells (Maurer and Fleischhauer 1979, Ramon y Cajal 1909−1911, Vaughan and Peters

1973). The horizontal spread of dendrites in layer I may extend from 1 to 2 mm (Colonnier 1964). Von Bonin and Mehler (1971) suggest that many of the connections between neighboring columns are mediated via layer I. Three-dimensional reconstructions of serial sections of layer I (Vaughan and Peters 1973, rat) reveal that the dendrites of pyramidal cells differ from those of nonpyramidal cells only by their spines. In single, thin sections only few spines can be found in continuity with their parent dendrite. Only about 50% of the spines contain a spine apparatus (Vaughan and Peters 1973) and can thus be discriminated unequivocally from small dendrites.

The light-microscopical staining of neural end-feet (Armstrong et al. 1965) demonstrates that there are differences of size, density, and distribution among the boutons within layer I (Colonnier 1967, cat). The superficial portion contains some large boutons terminaux. It is less densely populated with boutons than the profound portion. The axons projecting into layer I originate mainly outside the striate area, as has been shown with different methods in various experimental animals. Only a small number of them may be ascending axon collaterals of supragranular pyramidal cells. The following subcortical nuclei project into area 17: the corpus geniculatum laterale (Anker and Cragg 1974, cat; Carey et al. 1979, 1980, Tupaia, Galego; Garey and Powell 1971, cat, monkey; Lund 1973, *Macaca mulatta*; Peters and Feldman 1976, rat; Ribak and Peters 1975, rat; Schober and Winkelmann 1977, rat; Winfield and Powell 1976, cat), parts of the pulvinar (Benevento and Rezak 1976, Benevento et al. 1975, Carey et al. 1979, 1980: Tupaia, Galego; Ogren and Hendrickson 1977, *Macaca fascicularis*; Rezak and Benevento 1979, *Macaca mulatta*; Schober et al. 1976, rat), the claustrum, and the intralaminar nuclei of the thalamus (Carey et al. 1980). Cortical fibers come from the parastriate area (= area 18 Brodmann 1909; Benevento et al. 1975, *Macaca mulatta*; Rockland and Pandya 1979, *Macacus rhesus*; Tigges J et al. 1973, 1974, 1977, *Saimiri*; Wong-Riley 1979, *Macaca speciosa*) and from a temporal "association" area (termed MT = "middle temporal visual area" from Allman and Kaas 1971 for *Aotes trivirgatus*; Spatz 1977, *Callithrix*). Ferster and LeVay (1978, cat) applied the enzyme horseradish peroxidase to an incision into the optic radiation, close to the striate area, and found terminal arborizations, restricted to a very narrow band immediately below the external glial layer. The branches travel great distances (about 2 mm in one direction), but the number of dilations, probably boutons en passant, are relatively sparse. Ferster and LeVay believe that these axons either are, or resemble, geniculate afferents. They further assume that the geniculate axons do not change their diameter when they arrive in the cortex. Within the upper portion of layer I geniculate afferents have been shown to terminate (Ribak and Peters 1975, rat) with asymmetric synaptic contacts upon dendritic spines (Winfield and Powell 1976, cat). In the human striate area, within the sublamina tangentialis, the large boutons with spherical vesicles are probably geniculate terminals. Their morphological features closely resemble those of the large boutons of layer IVcβ (Braak E 1978; for comparison see also Tigges M et al. 1977, *Macaca mulatta, Saimiri*).

5 General Remarks Concerning Pyramidal and Nonpyramidal or Stellate Cells

Ramón y Cajal (1909–1911) describes two basic classes of Golgi-impregnated neurons in the isocortex: the pyramidal and the nonpyramidal neurons. Only the pyramidal cells reveal peculiarities which are characteristic of a defined layer, whereas nonpyramidal cells seem to lack them; at least we do not know of any at present.

It seems appropriate first to point out some general structural features common to the pyramidal cells of layers II–VI of the adult human striate area, and thus stress the morphological differences with respect to nonpyramidal cells. The description of the pyramidal cells of the various layers can then be confined to the layer-specific features.

5.1 Nissl Preparations

In Nissl preparations typical pyramidal cells are identified by the roughly triangular or pyramidal shape of their cell bodies. The origins of the apical and basal dendrites give the perikaryon a peaked appearance. Nissl substance continues from the soma into the proximal parts of the dendrites. The spherical nucleus has a smooth outline and contains small chromatin condensations and a nucleolus. In all optic levels of the section, the nucleus remains almost circular, only the nuclear diameter is changed. (In nonpyramidal cells the shape of the nucleus changes altering the optic levels of the section.) The nucleus is surrounded by a relatively narrow cytoplasmic rim (as compared with that of the large nonpyramidal cells). Either faintly tinged basophilic material is evenly distributed within the cytoplasm, or Nissl bodies occur adjacent to the somal periphery. The main characteristics are summarized in Table 2 and compared with those of the nonpyramidal cells, which are described in more detail in Sect. 11.2.

5.2 Golgi Preparations

The general morphological differences between pyramidal and nonpyramidal cells seen in Golgi impregnations are listed in Table 3. In general, the perikarya of pyramidal cells are more or less conical. Often they are not aligned in an exactly radial fashion and thus, seen in the projection, their shape is not always pyramidal in the section (Feldman and Peters 1978, rat).

The pyramidal cells exhibit polarity: they show an ascending apical dendrite and, opposite to it, a descending axon. Basal dendrites extend horizontally, or descend from the basal part of the soma. The descending axon gives rise to several collaterals. In general, when the axon becomes myelinated, the impregnation ceases. In an adult human, therefore, the axons of pyramidal cells cannot be traced upon entering the white matter, as they can in the early postnatal stages. The thick dendrites begin as gradually tapering processes. They give rise to several side branches, not only in their proximal parts but also in their distal parts. In the present study the side branches of the apical dendrite are called "apical side branches", those of the basal dendrites

10

Table 2. Morphological characteristics of the pyramidal and nonpyramidal cells in Nissl preparations

	Pyramidal cells	Nonpyramidal cells
Perikaryon	Peaked	Rounded
Nuclear-cytoplasmic ratio	Large	Variable
Position of the nucleus	Central	Eccentric
Contour of the nucleus	Spherical, smooth outline	Polymorphic with shallow to deep indentations
Distribution of chromatin	Few small and evenly distributed chromatin condensations; light appearance of the nucleus	Nucleus is more intensely stained; small chromatin condensations are near the nuclear envelope
Basophilic substance	Faintly tinged, either evenly distributed, or condensed in small Nissl bodies close to the soma membrane	Intensely stained and evenly distributed; Nissl bodies frequently fill up nuclear indentations
Stem of the dendrites	Basophilic material and Nissl bodies extend into the dendrites	Basophilic substance ceases with a crater shape in the soma and does not continue into the dendrites

"basal side branches". Distally from the ramification the dendrites undergo a reduction in their diameter. They display a slightly wrinkled outline and bear a moderate number of spines. Only the proximal segment of the dendrites, especially that of the apical dendrite (Schierhorn et al. 1973), lacks spines. This portion is considered to be an elongation of the soma; its length may be correlated with its phylogenetic evolution.

During early ontogenesis spine-free dendritic segments are not present (Lund et al. 1977, *Macaca nemestrina*). Classification of a neuron as a pyramidal cell is facilitated if the apical dendrite can be traced to layer I. This cannot be done in a number of neurons located in the deeper cortical layers, which fulfill the criteria mentioned above and therefore are classified as pyramidal cells (see also Lund 1973, Lund and Boothe 1975). Inverted pyramids (Van der Loos 1965) are rare.

For the morphological details of nonpyramidal neurons see Sect. 11.1.

In area 17 of the cat, most of the neurons can be classified physiologically as "simple", "complex", or "hypercomplex" cells (Hubel and Wiesel 1962). In an attempt to relate receptive field properties to neuronal morphology, Kelly and Van Essen (1974) and Lin et al. (1979) stained neurons with Procion yellow after classifying them as "simple", "complex", or "hypercomplex". The physiological and anatomical cell types cannot yet be correlated; all three physiological types could be either pyramidal or nonpyramidal cells.

5.3 Electron Microscopy

Generally the pyramidal cells have a triangular cell body. Frequently the apical dendrite can be traced for 20–50 μm. The nucleus is circular and contains a few small chro-

Table 3. Morphological characteristics of the pyramidal and nonpyramidal cells in Golgi preparations

	Pyramidal cells	Nonpyramidal cells
Shape of the perikaryon	Pyramid-shaped	Spherical, ovoid
Transition from the soma to the dendrite	Fluent	Abrupt
Dendritic arborization	From the soma arises an ascending apical dendrite, which gives off dichotomizing side branches and ends in terminal twigs; opposite the apical dendrite issue ramifying basal dendrites	Dendrites arise at arbitrary parts of the soma membrane and radiate in all directions
Mode of ramification	Dendrites also dichotomize in their distal parts	Dendrites branch infrequently, and usually near the soma
Diameter of the dendrites	Reduced as the dendrite extends distally; a considerable reduction occurs distally from the first ramification	Hardly reduced distally
Course of the dendrites	Within their main direction the dendrites take a fine, zig-zag course	The course of the dendrites appears straighter, the distances between the direction changes are greater than in the pyramidal cells
Outline of the dendrites	Appears slightly wrinkled	Appears smooth; irregularly arranged constrictions and dilations occur
Dendritic spines	Present except in the proximal dendritic segments	Absent
Axon	Arises at the base of the cell body or at a basal dendrite and takes a descending course	Arises at an arbitrary point of the soma and ramifies in the vicinity of the soma

matin condensations. It is located centrally, surrounded by a thin cytoplasmic rim which shows a moderate density of organelles (Peters and Kaiserman-Abramof 1970). The cisterns of the rough endoplasmic reticulum occur singly and are frequently branched. Lamellar bodies (Herndon 1964, LeBeux 1972) and subsurface cisterns (Rosenbluth 1962), as well as cisterns separated from the nuclear envelope by a plate of electron-dense, granular, ground substance (about 30 nm wide) are present. They seem to belong to the endoplasmic reticulum (Bestetti and Rossi 1980) and are considered to be involved in metabolic interchanges between the neuron and its environment (LeBeux 1972), e.g., between the nucleus and the cytoplasm and/or in stage-specific protein synthetic activity during development (Buschmann 1979). The Golgi complexes often occur opposite the origin of the dendrites. A varying number of lipofuscin granules (diameter 0.5–2.0 μm) are present only within the perikaryon. They do not usually penetrate into the processes (for exceptions see Braak E et al. 1980). The lipofuscin granules consist of an electron-dense matrix and electron-lucent droplets which bulge the enveloping membrane. Highly magnified micrographs reveal short, densely packed tubular subunits (diameter 9 nm) within the matrix (Braak E 1978). Axosomatic synaptic contacts are rare, and are always of the symmetric type.

Within the apical dendrites, polyribosomes and cisterns of the rough endoplasmic reticulum are located principally in the periphery, while the microtubules tend to occupy the more central parts of the dendrite. Synaptic contacts cannot usually be demonstrated along the proximal part of the apical dendritic shaft.

The organelles found in the soma, except the Golgi complexes, are also present in the proximal basal dendrites. Both symmetric and asymmetric synaptic contacts can occasionally be seen along the basal dendrites, close to the soma. The axon hillock and the initial segment do not differ in their morphology from those of the pyramidal cells in mammals (Palay et al. 1968, Peters et al. 1968). Synaptic contacts of both the symmetric and the asymmetric type are frequently present at this proximal axon segment. Subsurface cisterns are often located opposite an astrocytic profile (Reyners et al. 1977).

Different size classes of pyramidal cells can be distinguished in ultrathin sections if only those cells are taken into account which were cut through the middle of the spherical nucleus. This can easily be seen by the clear-cut appearance of the inner and outer nuclear membranes, which are at a distance from each other of about 50 nm. If a nucleus of 10 μm diameter is cut 1 μm above or below the equator, the two nuclear membranes appear superimposed.

The main morphological characteristics of the pyramidal cells are listed in Table 4. For a description of nonpyramidal neurons see Sect. 11.3.

Table 4. Electron-microscopical features of pyramidal and nonpyramidal cells

	Pyramidal cells	Nonpyramidal cells	
		Small	Large
Shape of the cell body	Triangular; the ascending apical dendrite can be traced for 20–50 μm	Polymorph; the origins of dendrites are rarely encountered	
Shape of the nucleus	Circular	Polymorph with shallow and deep indentations	
Amount of chromatin	Few small chromatin condensations, evenly distributed	Numerous small chromatin condensations, evenly distributed	
Number of organelles	Few	Few	Many
Distribution of organelles	Even	Even	Clusters of ribosomes, RER cisterns, and mitochondria are localized in the vicinity of the nuclear membrane
Number and types of axosomatic synapses	Small, symmetric	Small	Large; the asymmetric outnumber the symmetric
Number and types of synapses at the proximal dendritic segments	Apical dendrite: rare Basal dendrites: several asymmetric and symmetric	At all dendrites several asymmetric and symmetric contacts	

6 Pyramidal Cells and Neuropil of Layer II

6.1 Nissl Preparations

Layer II is densely populated with small neurons, and its lower margin is difficult to determine (Fig. 10). Most of the neurons are small pyramidal cells. Intermingled with both nonpyramidal and glial cells, they tend to aggregate in clusters, triads, horizontal rows, and vertical columns (Fig. 11a–c). The narrow cytoplasmic rim of the small pyramidal cell (diameter 8–11 μm) is faintly tinged. At the origin of the apical dendrite a more pronounced basophilic area is visible (Fig. 11a, b). The apical dendrite (diameter 2.5 μm) can be traced for about 15 μm. Often both the perikaryon and the apical dendrite are associated with a satellite glial cell (Fig. 11a, b; see also Von Economo and Koskinas 1925).

6.2 Golgi Preparations

In Golgi impregnations two types of small pyramidal cells can be distinguished: (a) Py II/1 small pyramidal cells, and (b) Py II/2 small pyramidal cells with an uncommon dendritic arborization.

Py II/1: Small Pyramidal Cells

The apical dendrite of these small pyramidal cells gives rise to a few side branches, most of which do not bifurcate. The proximal side branches take a more horizontal course (Fig. 12a), while the distal ones ascend obliquely. The terminal tuft of the apical dendrites originates in the vicinity of the layer I/II border. Distally from the spineless segment (see Sect. 5.2., Schierhorn et al. 1973) the apical dendrite and its side branches are covered with sessile and stalked spines (Peters and Kaiserman-Abramof 1970).

The basal dendrites bifurcate repeatedly and take either a horizontal or a descending course. Additional dendritic stems occasionally arise from an upper part of the soma, and their side branches take an ascending course (Fig. 12b). The diameter of the basal dendritic tree exceeds that of the apical dendritic tree. In general the basal dendritic stems are spineless, but their side branches are covered with sessile and stalked spines. On examining tangential sections of layer II, it becomes evident that the basal dendritic stems originate at the largest possible distance from each other. Sometimes the descending axon – giving rise in its course to several side branches (Fig. 12b) – can be traced for about 120 μm. Occasionally a pyramidal cell (Fig. 12c, d) which lacks spines on its apical dendritic tree occurs next to a spined one. In one case the axon can be traced to layer IVa. Some collaterals arise at right angles from its proximal 200 μm and take an obliquely ascending course. Within layer IIIab the axon is covered with appendages ending in a bulb. It is difficult to decide whether these neurons are merely insufficiently impregnated or represent a new class of pyramidal cells. Pyramidal cells with sparsely spined dendrites have been reported by Ramón y Cajal (1909–1911, Fig. 351 cell A) and Braak H (1976).

14

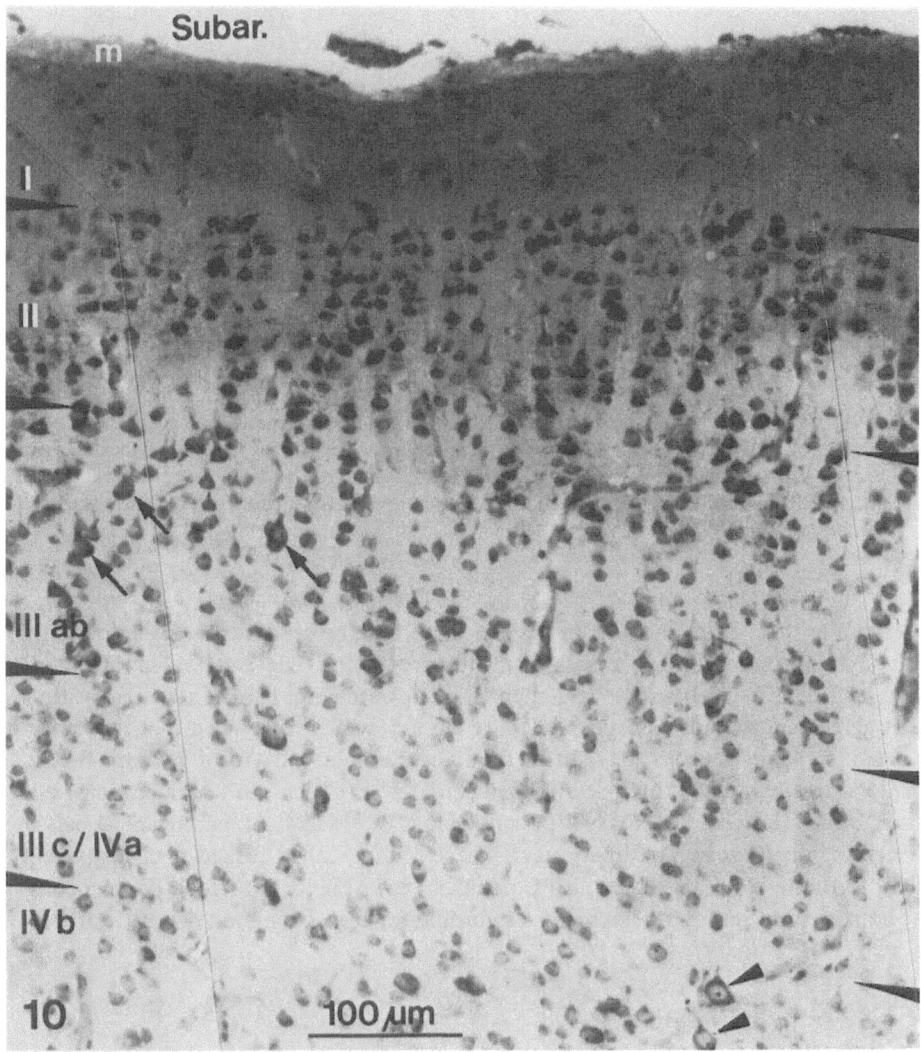

Fig. 10. Nissl-stained section through the human striate area (Araldite, 10 μm, methylene blue; 66-year-old man) showing the upper three layers and the upper tier of layer IVb. The cortex borders with the sharply outlined membrana limitans gliae superficialis (*m*) on the subarachnoid space (*Subar.*). The lower border of the densely populated layer II can be drawn where the cell density is reduced. Layer IIIab is characterized by some large pyramidal cells (*arrows*). The appearance of a large number of small, faintly tinged nerve cells marks the beginning of layer IIIc/IVa. In layer IVb the cell density decreases and some large solitary Meynert-Cajal cells are present (*arrowheads*)

Py II/2: Small Pyramidal Cells with Uncommon Dendritic Arborization

The appearance of these cells depends on their location within a gyrus. Especially at the summit of the gyrus, spinous neurons which lack a typical apical dendrite can be seen. The basal dendrites, in contrast, show no irregularities. Either one thick stem emerges and soon bifurcates (Fig. 13a, e), or two thick branching dendrites issue

15

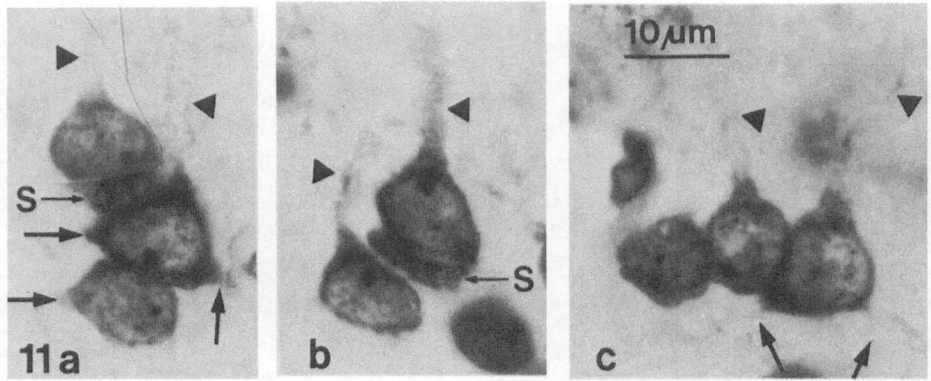

Fig. 11a–c. Nissl-stained layer-II pyramidal cells (Araldite, 10 μm, methylene blue; 66-year-old man). They are arranged in vertical columns (*a, b*) or in horizontal rows (*c*) and are associated with satellite cells (*S*). Note the homogeneously tinged cytoplasmic rim and the basophilic substance, which marks the apical dendrite (*arrowheads*) and also extends for a short distance into the basal dendrites (*arrows*)

directly (Fig. 13c, d) from the tapering apical part of the cell body. In other cases the cell body appears tilted, and a thick dendrite arises from the lateral portion of the cell body; it gives off side branches and then turns upward (Fig. 13b). These observations are in agreement with earlier morphological studies, which recognized that the dendritic arbor of some neurons may differ from the common pattern of pyramidal cells: The dendrites may issue in a pattern similar to the one found in nonpyramidal cells: star pyramids (Lorente de Nó 1938, Szentágothai 1971) or the tapering apical part of the neuron may already bifurcate near the cell body, called forked neurons (Gabelzellen) (Kirsche et al. 1973; see also Colonnier 1967, Fig. 3 top right). Kirsche considers impairments during development responsible for the uncommon dendritic arbor.

6.3 Electron Microscopy

Small Pyramidal Cells

The layer-II pyramidal cells have a narrow cytoplasmic rim where a few single cisterns of the rough endoplasmic reticulum are arranged parallel to the nuclear outline (Fig. 14). Sometimes two perikarya lie in close apposition without an intervening glial process. The two adjacent membranes are connected by puncta adhaerentia. The soma membranes rarely receive symmetric synaptic contacts. Some of these terminals may originate in nonpyramidal neurons within the layer (Peters and Fairén 1978, rat).

Fig. 12a–d. Golgi-impregnated small pyramidal cells of layer II (frozen section, 100 μm, Permount, 72-year-old man). *Ax*, axon. The border of layer I/II is marked. *a, b* Pyramidal cells, the apical dendrite and side branches of which bear a moderate number of spines (see also the middle of the three cells in *c*). *Arrowheads* point to axon collaterals. *c, d* Pyramidal cells of which the apical den-

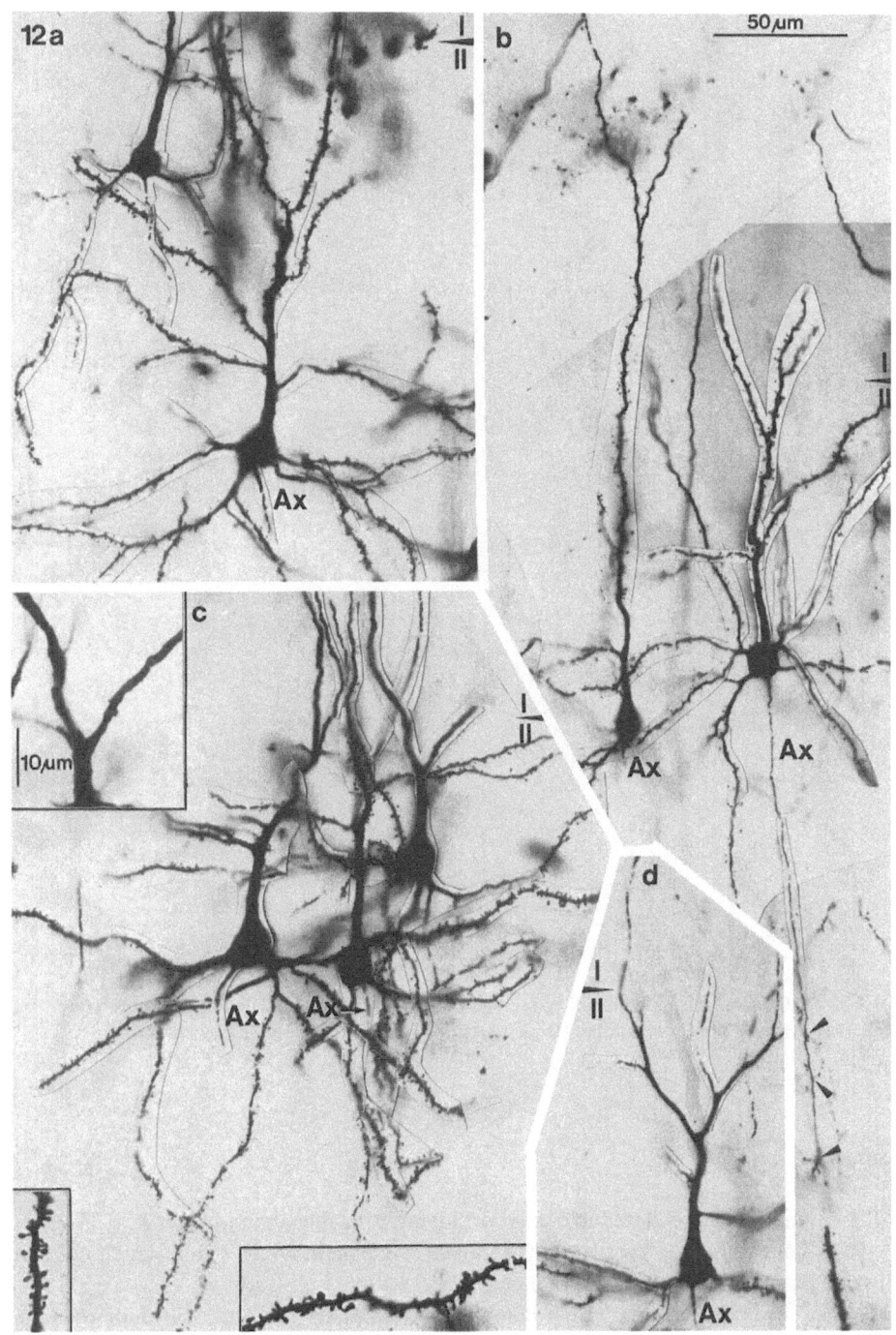

drite is spineless (see inset in *c* in the top left corner; exception, the middle cell in *c*). Their apical side branches are sparsely spined and the basal side branches are studded with stalked spines (insets at the bottom of *c*). The scale is suitable for all the insets. The scale of *b* is also appropriate for *a, c,* and *d*

17

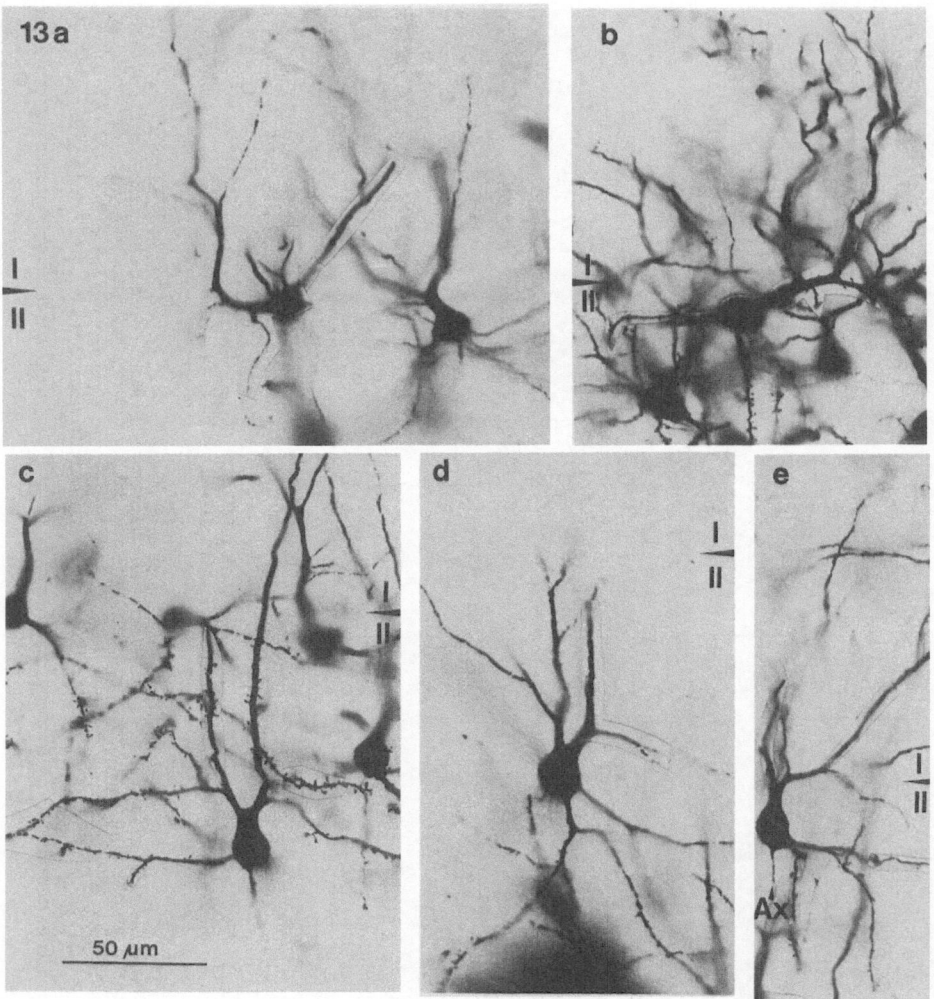

Figs. 13a–e. Golgi-impregnated small pyramidal cells with uncommon dendritic arborization. *Ax*, axon. The layer I/II border is marked (frozen section, 100 μm, Permount; 72-year-old man)

Neuropil

(a) The number of myelinated axons is considerably smaller than in layer I.

(b) The number of profiles having diameters smaller than 1 μm, which contain only one or two microtubules, is reduced.

(c) A few boutons containing dense-core vesicles are present. Still numerous are small boutons with spherical vesicles, which make asymmetric synaptic contacts at various dendritic profiles; several boutons with flat vesicles occur. Autoradiographic light-microscopical investigations reveal that the lower half of layer II receives an input from the lateral and inferior pulvinar (Ogren and Hendrickson 1977, *Macaca fascicularis*).

18

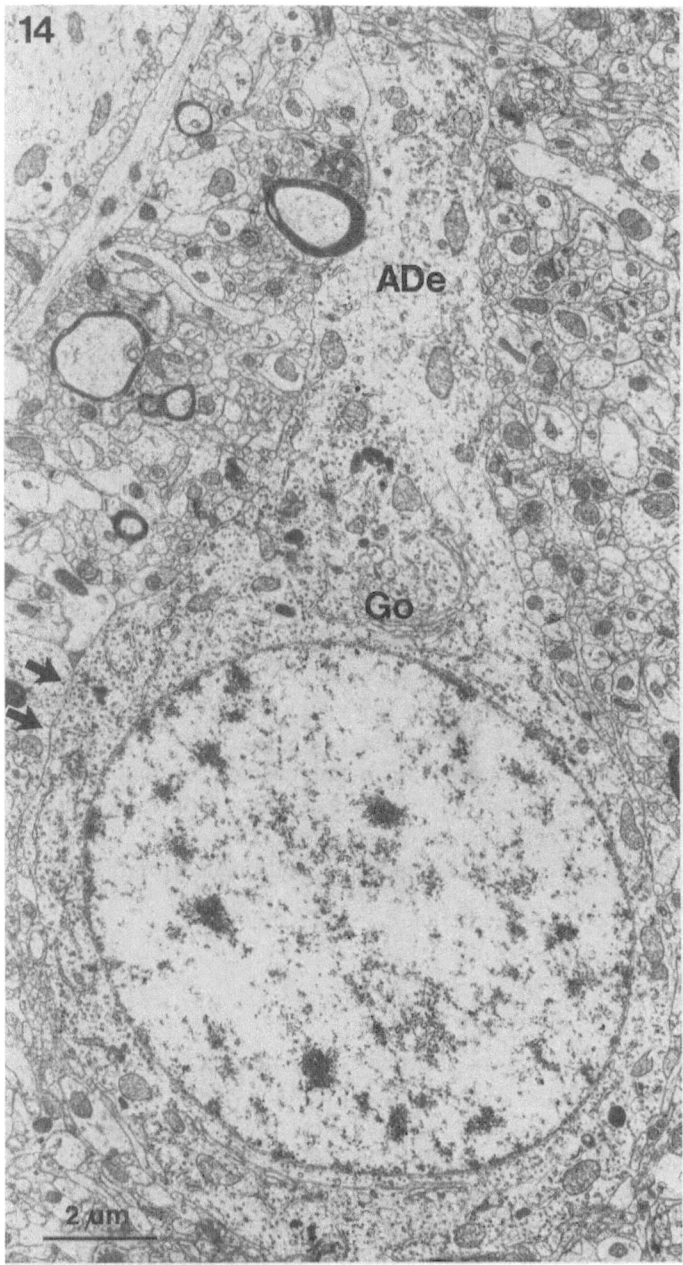

Fig. 14. Electron micrograph of a small layer-II pyramidal cell which is in close apposition with another neuron (*arrows*). Near the origin of the apical dendrite (*ADe*) a Golgi complex (*Go*) is located (66-year-old woman)

(d) The profiles of apical dendrites (diameter 2—4 μm) are frequently encountered.

(e) Radially oriented, dendritic (?) profiles (diameter 1.5 μm) can be traced up to 40 μm; their mitochondria appear to be somewhat darker than those of the apical dendrites, and they receive several asymmetric synaptic contacts.

(f) The number of profiles which contain glial filaments and glycogen particles, and thus belong to fibrillary astrocytes, are less often seen, while more and more protoplasmic astrocytic processes appear.

7 Pyramidal Cells and Neuropil of Layer IIIab

7.1 Nissl Preparations

Layer IIIab contains small and medium-sized pyramidal cells and nonpyramidal cells. Its lower margin is defined by an almost abrupt increase in the number of small cell bodies (Fig. 10). The small pyramidal cells of layer IIIab look like those of layer II (Fig. 15a—d), whereas the medium-sized ones (diameter 12—15 μm; Fig. 15e—g) have a larger nucleus and a broader cytoplasmic rim which contains more intensely stained basophilic material. Frequently a "basophilic cap" is near the origin of the apical dendrite. The nuclei of the human striate area pyramidal cells are smoothly contoured and spherical. In this point they differ from those of animals, which are indented (Garey 1971, monkey, cat, visual cortex; Jones and Powell 1970, cat, somatosensory cortex; Parnavelas et al. 1977b, rat, visual cortex; Peters and Kaiserman-Abramof 1970, rat, parietal cortex).

7.2 Golgi Preparations

In Golgi impregnations within layer IIIab two size-classes of pyramidal cells are present: (a) Py IIIab/1 small pyramidal cells, and (b) Py IIIab/2 medium-sized pyramidal cells.

The medium-sized IIIab pyramidal cells are usually regarded as the most typical (Parnavelas et al. 1977a, b), but Kirsche et al. (1973, somato-sensory cortex, rat) also call attention to the existence of small ones.

Py IIIab/1: Small Pyramidal Cells

The small pyramidal cell gives rise to a fine apical dendrite; its impregnation ceases in the vicinity of the layer I/II border and no terminal tuft is visible (Fig. 16a—d). The basal dendrites radiate for about 50 μm. Both apical and basal side branches are studded with stalked spines. In several cases the descending axon is traced for about 150 μm (Fig. 16a—c), giving off ascending and repeatedly ramifying collaterals.

These small pyramidal cells resemble identically named nerve cells shown in monkeys (Lund et al. 1977, *Macaca nemestrina*; Tömböl 1978, *Macaca rhesus*).

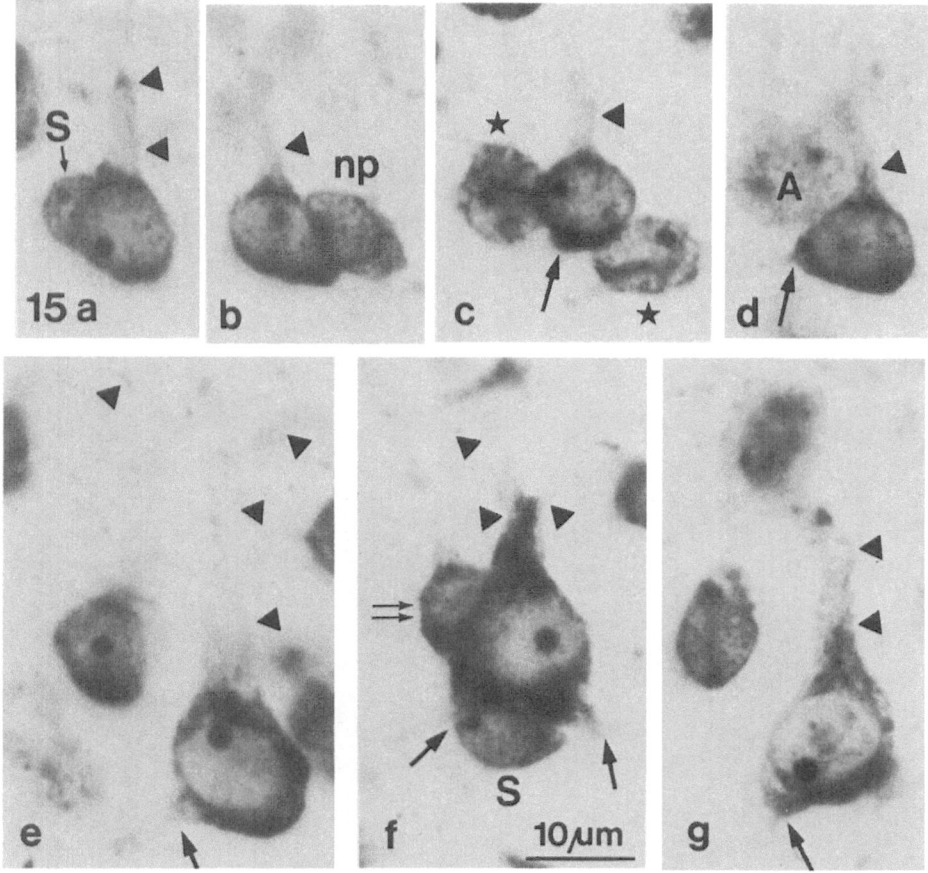

Fig. 15a–g. Examples of Nissl-stained small (*a–d*) and medium-sized (*e–g*) pyramidal cells of layer IIIab (Araldite, 10 μm, methylene blue; 66-year-old man). The proximal parts of the apical dendrites (*arrowheads*) contain small amounts of basophilic substance. Of the basal dendrites (*arrows*), only the origins are visible. Frequently, the pyramidal cells are associated with various types of cells, for example in *a* and *f* with satellite cells (*S*), in *b* with small nonpyramidal cells (*np*), in *c* with small, heavily pigmented nonpyramidal cells (*asterisk*), in *d* with astrocytes (*A*) and in *f* with another pyramidal cell (*double arrow*)

Py IIIab/2: Medium-sized Pyramidal Cells

The medium-sized neurons show the most typical morphological features of pyramidal cells (Fig. 17a–d). The proximal part (about 60 μm) of the stout apical dendrite (diameter 4–6 μm) gives rise to a few side branches which extend for about 100–150 μm. The part (about 40 μm) of the apical dendrite joining the spine-free segment is covered with stalked spines, whereas the terminal tuft only rarely bears spines. The distal apical side branches take a less oblique course than the more proximal ones. Side branches originating in the spinefree segment of the apical dendrite are spinous from their origin.

The basal side branches radiate for about 150 μm into the deeper layers. They are covered with stalked spines slightly less densely than are the apical side branches. The

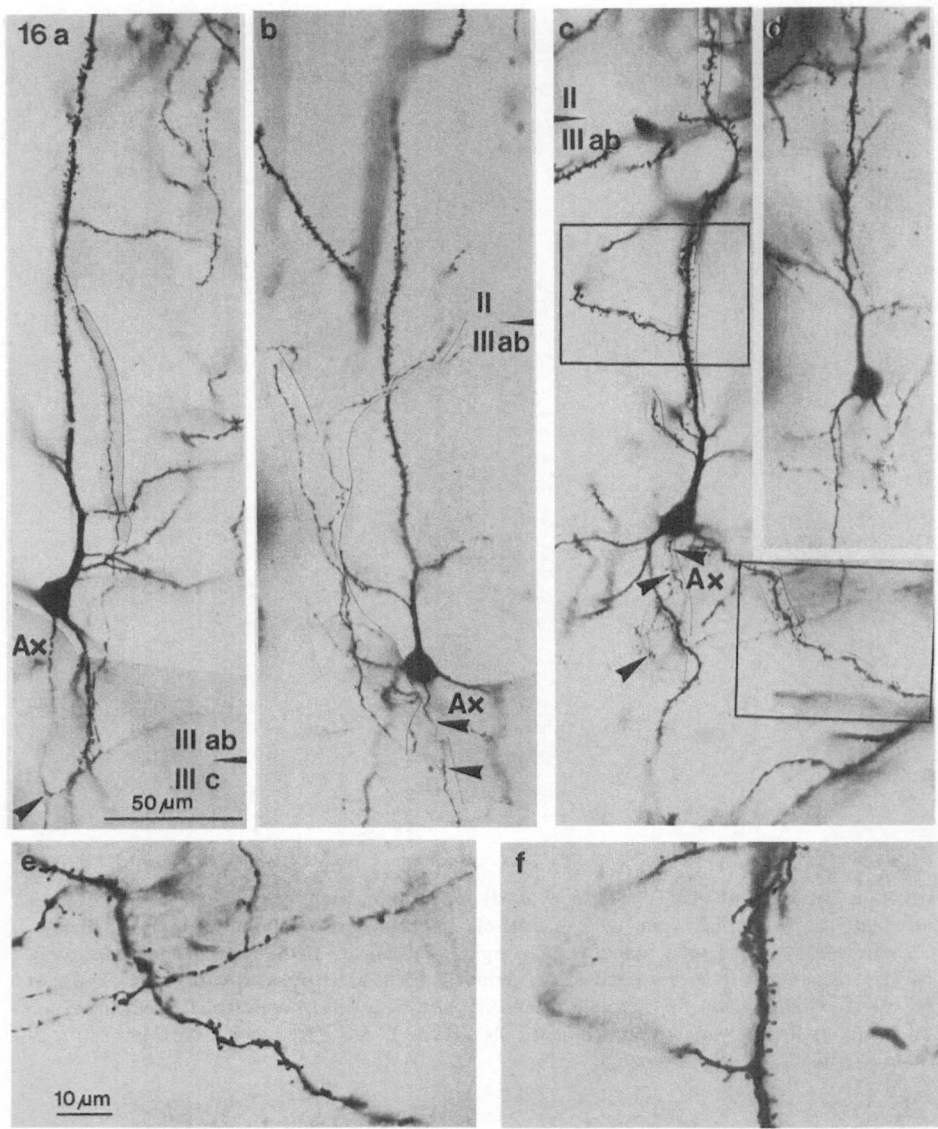

Fig. 16a–f. Photomicrographs showing in *a–d* small pyramidal cells of layer IIIab after Golgi impregnation (frozen section, 100 μm; Permount; 72-year-old man). Unbranched side branches already arise from the proximal parts of the apical dendrites. Both the apical dendrite and the apical side branches are covered with a moderate number of stalked spines. Shown in *f* is a higher magnification of the framed area of the apical dendrite of *c*. Only a few basal side branches are present, also bearing a moderate number of spines as is demonstrated in *e* at a higher magnification of the framed area of *c*. In the examples of *a–c* the descending axon (*Ax*) gives rise to recurrent ascending collaterals (*arrowheads*). The approximate border of the layers is indicated

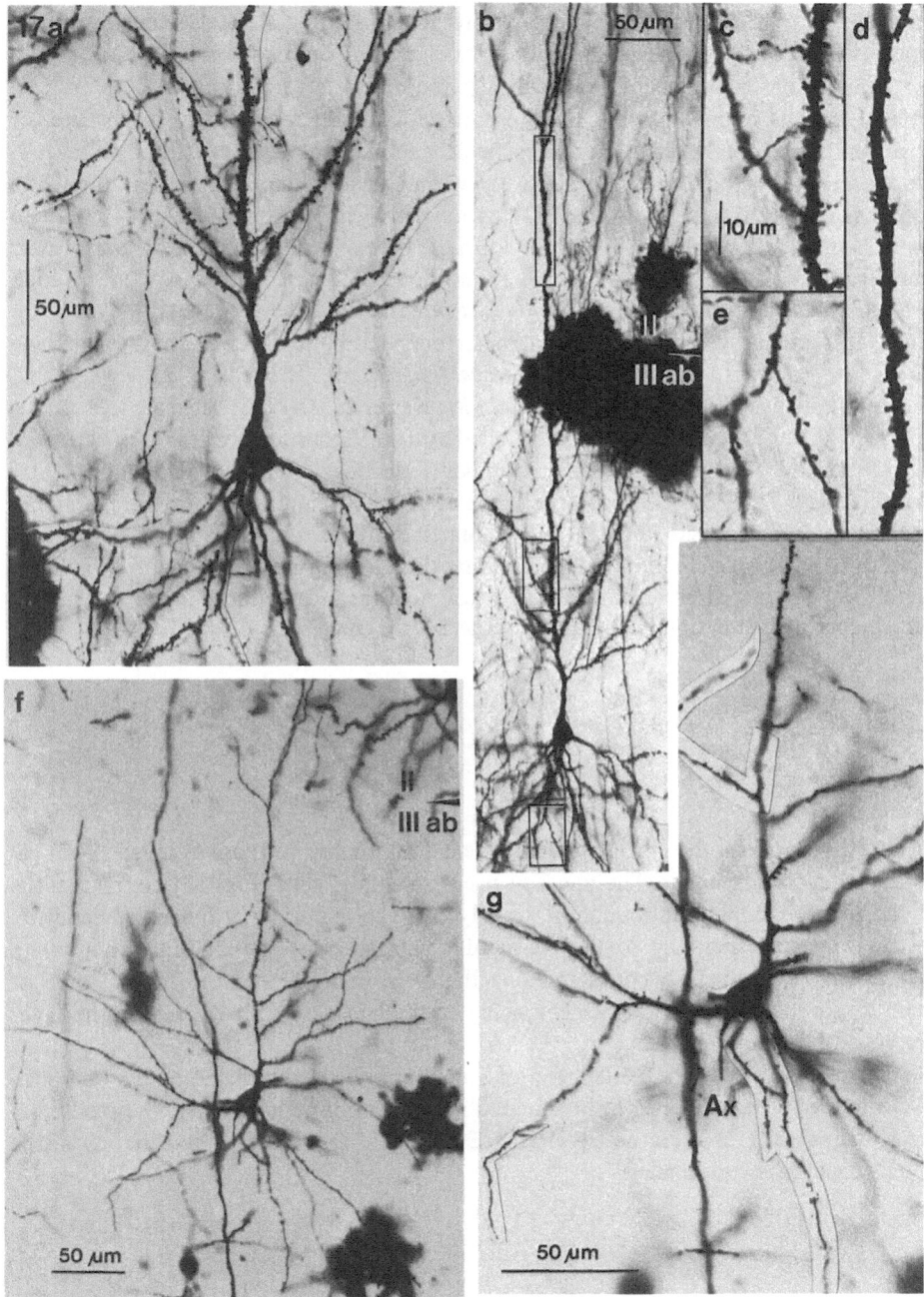

Fig. 17a–g. Micrographs of Golgi-impregnated medium-sized pyramidal cells (frozen section, 100 μm, Permount; 72-year-old man). The borders of the layers are indicated. The entire cell of *a* is shown in *b* at lower magnification. The framed areas of *b* correspond to higher magnified photographs of the apical dendrite in *c* and *d* and to the basal dendrites in *e*. Note the different spine densities in the various dendritic segments. The entire cell of *g* is shown in *f* at a lower magnification

23

axon can be traced down to layer IVa, there giving rise to collaterals which ascend or remain within the layer. The collaterals end with a small bulb.

Intracellular injection of horseradish peroxidase (Gilbert and Wiesel 1979, cat) reveals that the axon of layers II/III pyramidal cells branches richly in its layer of origin and exceeds the cell's dendritic field. The axon collaterals extend to layer I. The descending axon projects extensively within layer V and then enters the white matter (see also Allman and Kaas 1971, *Aotes trivirgatus*; Allman et al. 1973, *Galago senegalensis*; Doty et al. 1964, *Saimiri sciureus*; Spatz 1975, *Callithrix jacchus*; Spatz and Tigges 1972, *Callithrix jacchus*; Spatz et al. 1970, *Saimiri sciureus*; Woolsey et al. 1955, *Callithrix jacchus*; Zeki 1971, *Macaca mulatta*). Hence, the layers II/IIIab pyramidal cells mainly establish an interlaminar connection to layer V (Levey and Jane 1975, rat; Lund and Boothe 1975, *Macaca mulatta, M. fascicularis, M. nemestrina;* Martinez-Millán and Holländer 1975, *Saimiri*; Nauta et al. 1973, rat, cat, tree shrew; Spatz et al. 1970, *Callithrix*). In addition they form a cortico-cortical connection to area 18 (Lund et al. 1975, macaque monkey; Martinez-Millán and Holländer 1975, Rockland and Pandya 1979, *Macaca*; Spatz et al. 1970), and to the "middle temporal visual area" (Gilbert and Kelly 1975, cat; Martinez-Millán and Holländer 1975). Layer IIIab pyramidal cells are likely to receive direct input from layer $IVc\beta$ neurons and can therefore be regarded as second-order cortical neurons in the processing of parvocellular geniculate information (Lund and Boothe 1975).

7.3 Electron Microscopy

Pyramidal Cells

In ultrathin sections the small pyramidal cells look similar to those of layer II, but generally they occur singly. The medium-sized pyramidal cells (Fig. 18) contain stacks of the rough endoplasmic reticulum, usually close to the origin of the apical dendrite, in this way corresponding to the "basophilic cap" in Nissl preparations. There are a few axosomatic synaptic contacts along the large perimeter of the soma. Some of them could be axon collaterals belonging to stellate cells located within the layer (Peters and Fairén 1978, rat).

On rare occasions the perikaryon of a medium-sized pyramidal cell may be partly covered by a myelin sheath. Myelinated nerve cell bodies have been described in several brain areas (for literature see Braak E et al. 1977). Thus far, the function of the myelin sheath remains unclear.

Neuropil

Among the organelle-rich apical dendrites a relatively large number of radially oriented dendritic profiles (diameter 1.5 μm) can be traced for about 80 μm. A new type

---

Fig. 18. Electron micrograph of a medium-sized layer-IIIab pyramidal cell (66-year-old woman). The cytoplasmic organelles are evenly distributed within the narrow cytoplasmic rim. Within the apical dendrite (*ADe*) RER cisterns, ribosomes and Golgi complexes (*arrowhead*) accumulate, whereas other parts contain only microtubules. Note the smoothly contoured nucleus. An axo-

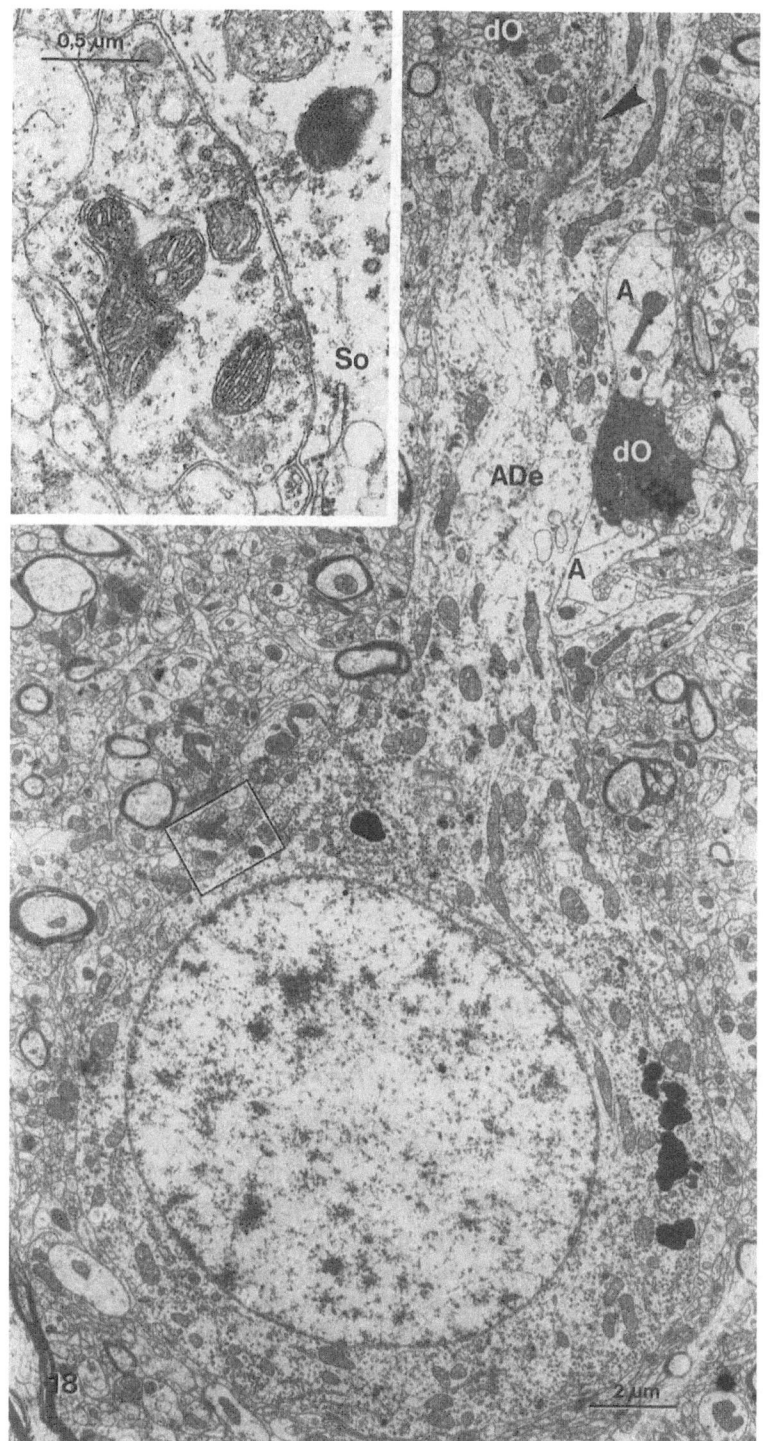

somatic Gray's type-2 synaptic contact is present within the *framed area* and shown at a higher magnification in the *inset*. Profiles of astrocytic processes (*A*) and dark oligodendrocytes (*dO*). *Inset*: *So*, soma

of dendritic (?) profile becomes visible, which is also oriented in a radial fashion and appears twisted; mitochondria with many cristae and an electron-dense matrix tend to aggregate into clusters; these profiles are frequently lined by boutons with spherical vesicles making multiple asymmetric synaptic contacts. Similar profiles are present in layer IVcβ (Figs. 8 and 9 in Braak E 1978). Slender profiles (diameter 0.2–0.6 μm), which contain a few microtubules, are cut longitudinally for about 5 μm. Myelinated axons appear in vertical bundles. The claustrum (Carey et al. 1980, *Tupaia*) and parts of the pulvinar (Rezak and Benevento 1979, *Macaca*), as well as area 18 (Brodmann), (Wong-Riley 1978, *Saimiri*) appear to provide input to layers II/IIIab. The projections coming from the pulvinar occur in patches. Presumably they are represented by terminals containing round vesicles with asymmetrical contacts on dendritic spines (Rezak and Benevento 1979).

8 Pyramidal and Polygonal Neurons and Neuropil of Layers IIIc/IVa, IVb, IVcα, and IVcβ

8.1 General Remarks

These layers display different packing densities of their cell bodies (Fig. 19) but they have one feature in common: they are composed of neurons, most of which cannot be categorized as true pyramidal or nonpyramidal cells. With regard to the shape of their cell body they are here designated as "polygonal neurons". These cells have frequently been referred to as "spiny stellate cells" (Garey 1971, LeVay 1973, Lund 1973), or simply as "stellate cells" (Ramón y Cajal 1909–1911, Valverde 1971). Lorente de Nó (1938) named similar neurons "star pyramids"; this term was also adopted by Jones (1975, *Saimiri*, somato-sensory cortex) for his type 7 "non-pyramidal cells". But in this context it should be mentioned that Szentágothai (1971, cat, monkey, somatosensory cortex) also used the term "star pyramid" for layer-II neurons, since their apical dendrites bifurcate in the immediate vicinity of the soma (Figs. 13c–e).

The polygonal neurons have some morphological features of both the pyramidal and nonpyramidal cells in common. On closer examination the polygonal nerve cells appear to be considerably modified pyramidal cells, rather than altered nonpyramidal cells (see Sect. 5). In Nissl preparations (Fig. 20a–o) the spherical nucleus of polygonal nerve cells is lightly stained and has a smooth outline; it contains a few small chromatin condensations. Basophilic substance is present in peripheral parts of the soma and the proximal dendritic segments as it is in typical pyramidal cells. Golgi impregnations display additional features which are usually found in pyramidal cells: the gradually cone-shaped transition from the soma to the dendrites, the presence of spines along the dendrites, and the dendrites which also dichotomize in their distal parts. Furthermore, the dendritic diameter is considerably reduced from the proximal to the distal part. The axon normally arises from the basal parts of the soma and projects toward the white matter. The polygonal cells lack a typical apical dendrite. Moreover, the dendrites may emerge at any point of the soma. It is mainly the uncommon dendritic arbor which does not permit classification of the polygonal neurons with real pyramidal cells. Layer IV neurons presumably undergo a transformation from a

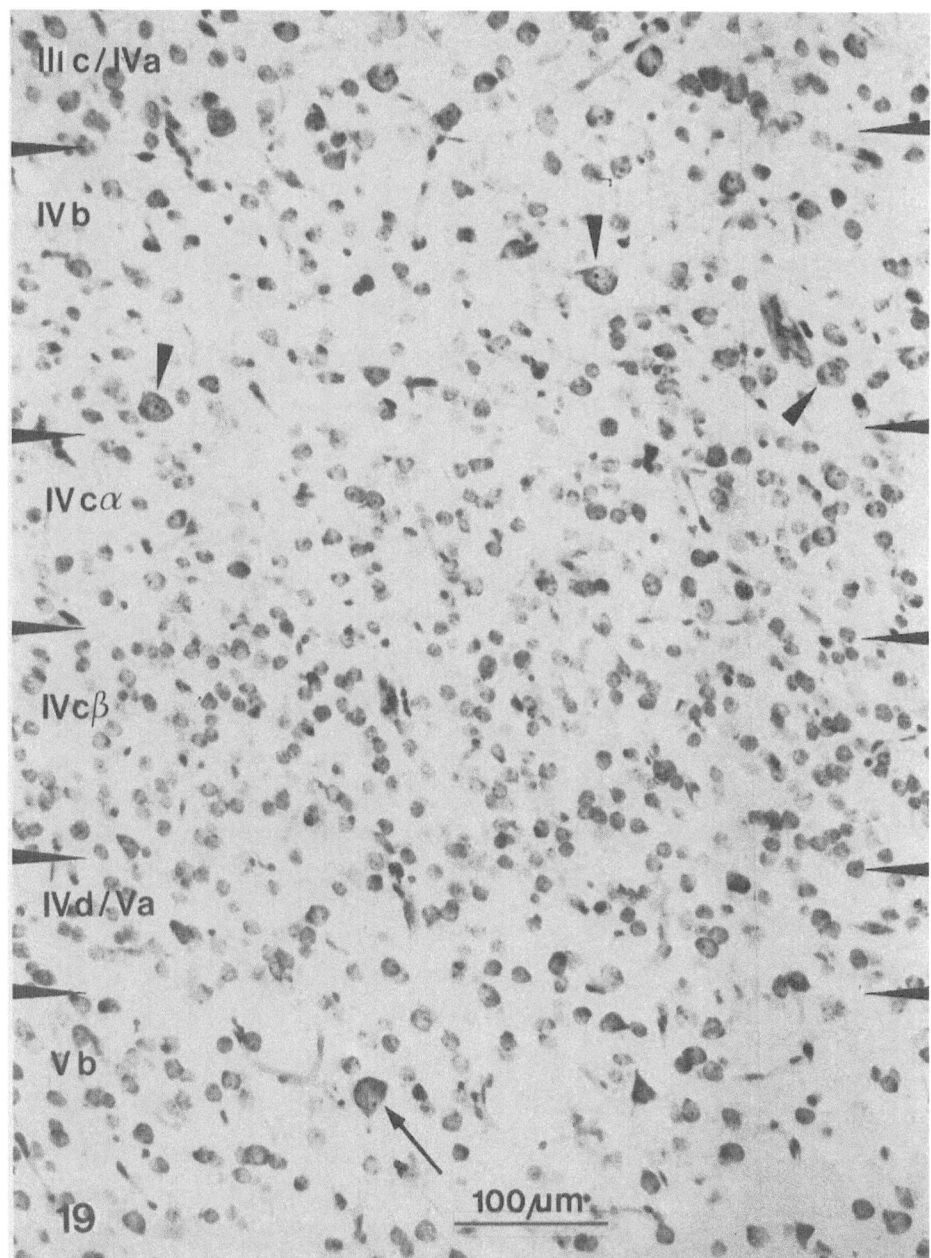

Fig. 19. Nissl-stained section through the human striate area showing the layer IV (Araldite, 10 μm, methylene blue; 66-year-old woman). Layer IVb contains few cell bodies; among them are large Meynert-Cajal solitary neurons (*arrowheads*). In lamina IVcα the number of cell bodies has increased. Lamina IVcβ is densely populated by one type of small, faintly tinged nerve cell which tends to lie in clusters. Lamina IVd/Va contains fewer nerve cells and they vary in size. In lamina Vb the number of cell bodies has again decreased; Vb contains the largest nerve cell bodies of the striate area, the Meynert pyramidal cells (*arrow*)

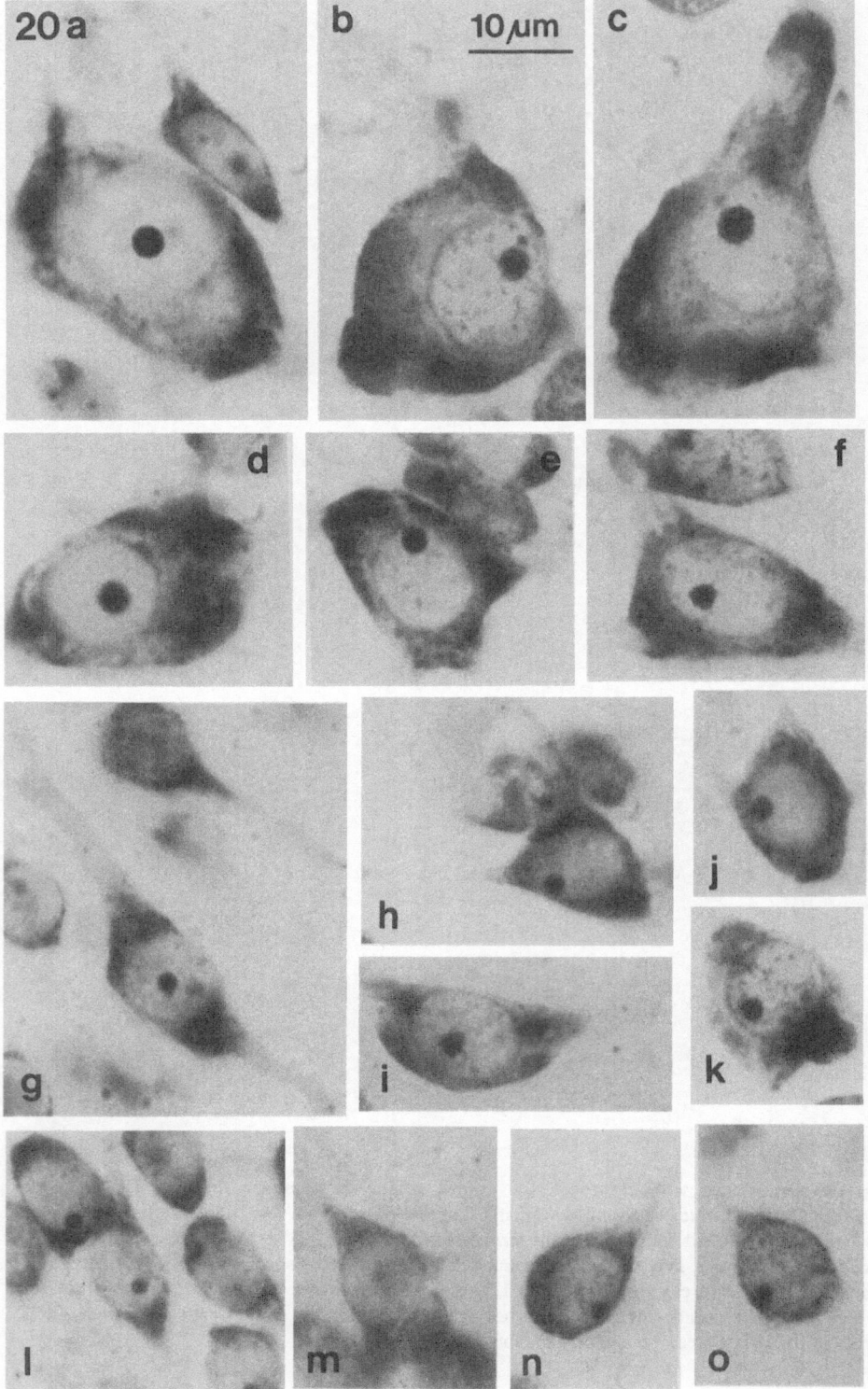

20 a   b   10 μm   c

d   e   f

g   h   j

i   k

l   m   n   o

pyramidal to a polygonal shape. Along the upper and lower borders of layer IV, i.e., layers IIIc/IVa and IVd/Va, pyramidal cells are still present (Fig. 23a) and are intermingled with the polygonal neurons. Probable transitional stages ranging from a typical pyramidal cell shape to a polygonal one can be observed especially in layer IIIc/IVa.

Figure 21b—d demonstrates pyramid-shaped cell bodies giving rise to an apical process which ramifies close to the soma (40 μm); the branches take an oblique course. Often two ramifying dendrites also arise (Fig. 21b) from the tapering apical part of the soma. In other cases, the dendrites emerge only from the lateral parts of the cell body, the upper part being smooth (Fig. 22d) or occasionally invested with a stubby process (Fig. 22a, lower cell body). The basal part of the cell body seems to be reserved for the axon hillock; thus the long axis of these neurons lies horizontal. Large neurons of this type have already been described by Meynert (1867, 1868) and Ramón y Cajal (1900, 1909—1911), and are frequently referred to as the Meynert-Cajal solitary neurons. In layers IVd/Va and Vb polygonal neurons become rare. Pyramidal cells predominate, although many of them show only a delicate apical dendrite which does not usually reach the molecular layer.

8.2 Nissl Preparations

The various layers can be distinguished from each other by differences in the packing density and size classes of the neuronal perikarya (Fig. 19). In comparison with layer IIIab, layer IIIc/IVa is a band of closely packed, small and medium-sized neurons, among which some larger ones are interspersed. Layer IVb is a band of less densely packed neurons. It harbors the large solitary Meynert-Cajal neurons. In fiber preparations (Braak H 1980) this layer contains the stria of Gennari. Layer IVcα is densely populated by medium-sized and small polygonal neurons. The medium-sized neurons are stained in varying intensity. Many of the small neurons are associated with satellite glial cells. The adjacent layer IVcβ is almost exclusively composed of small, densely packed, and faintly tinged polygonal neurons. These small neurons are characterized by numerous lipofuscin granules (Braak H 1976, Braak E 1978) which make this layer especially prominent in pigment preparations (Braak H 1977).

Fig. 20a—o. Nissl-stained polygonal neurons (Araldite, 10 μm, methylene blue; from two 66-year-olds, a man and a woman). Note the smooth outline of the spherical nucleus. Faintly tinged basophilic substance extends into the proximal segments. In $a-c$ are examples of the large polygonal neurons, the Meynert-Cajal solitary neurons, of layers IIIc/IVa (in a and b) and IVb (in c); in $d-f$ examples of the medium-sized polygonal neurons of layers IIIc/IVa (in d and e) and IVb (in f) are shown. Intensely stained basophilic substance tends to accumulate in the more peripheral parts of the large and medium-sized cell bodies.
In $g-k$ are examples of small polygonal neurons of layers IIIc/IVa (in g and h), IVb (in i and j), and IVcα (in k), which have only a narrow cytoplasmic rim and more evenly distributed Nissl substance.
In $l-o$ are examples of small polygonal neurons having very little cytoplasm from layers IIIc/IVa (in l), IVb (in m), IVcα (in n), and IVcβ (in o)

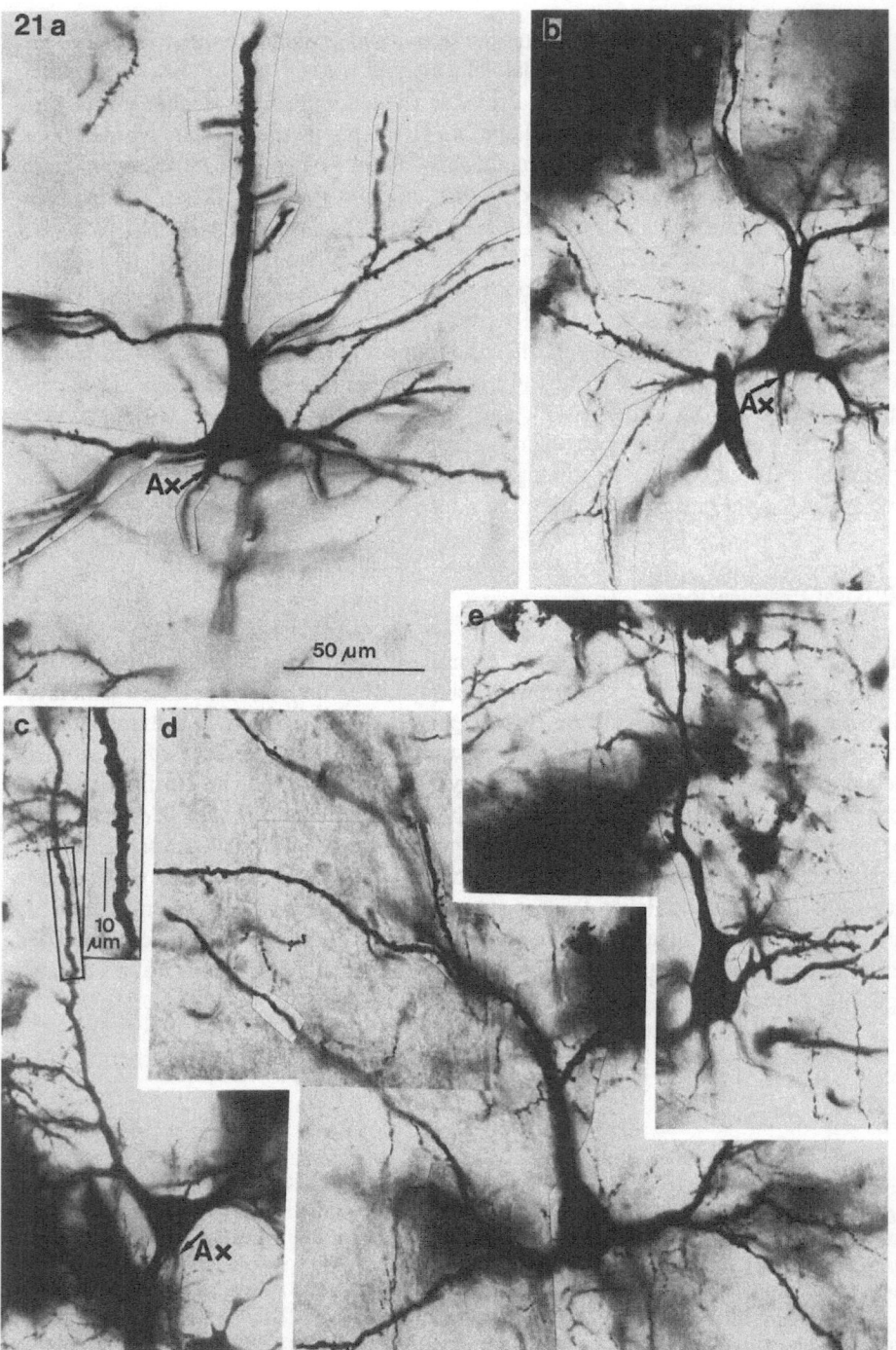

Fig. 21a–e. Photomicrograph of Golgi-impregnated large polygonal neurons (*b–e*: Poly-n IIIc/IVa–IVb/1) and one pyramidal neuron (*a*: Py IIIc/IVa/1) of layer IIIc/IVa (frozen section, 100 μm, 72-year-old man). They exhibit features which allow the hypothesis that they are derived from pyramidal cells. Example *a* is regarded as a real pyramidal cell, the apical dendrite of which goes beyond the plane of the section. The conspicuous alteration occurs at the apical dendrite. In *b*

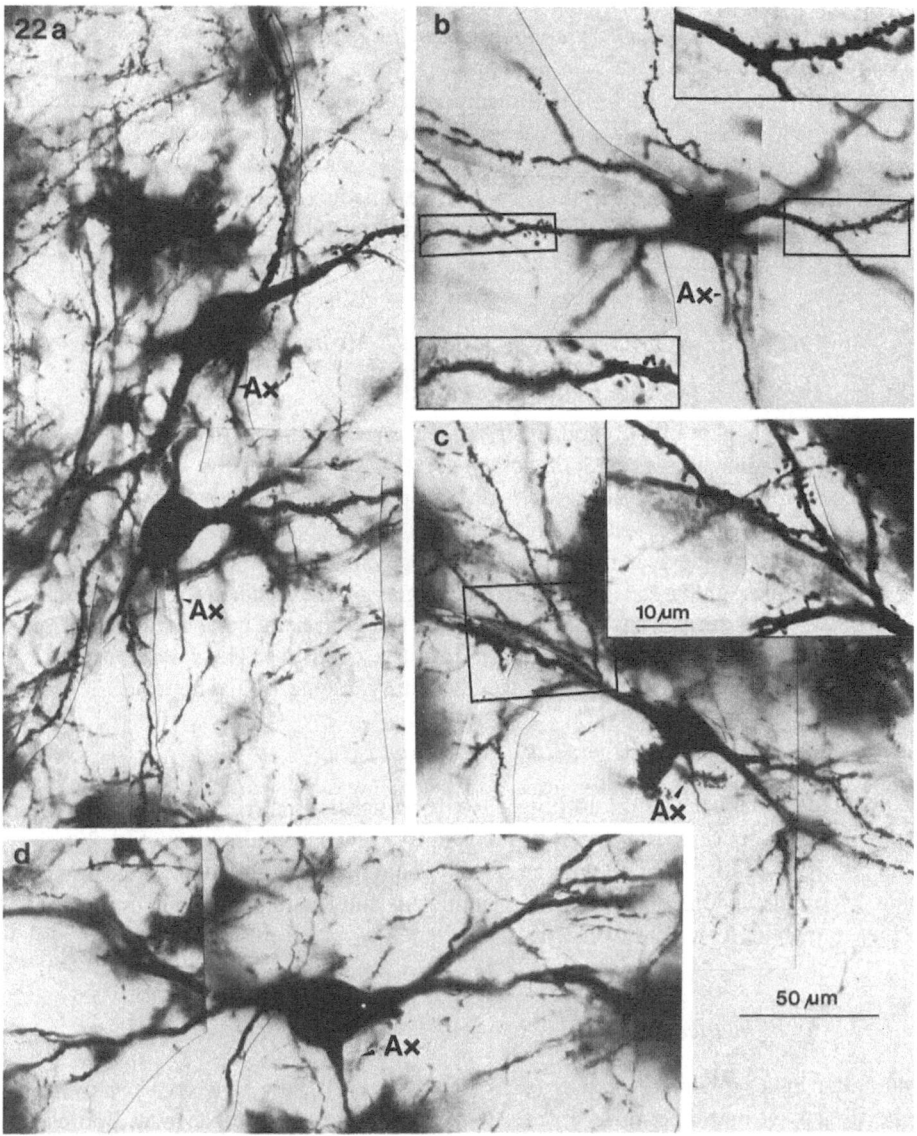

Fig. 22a–d. Photomicrographs of Golgi-impregnated Meynert-Cajal solitary neurons, the large polygonal neurons of layer IIIc/IVa in *a* and *b*, of layer IVb in *c* and *d* (frozen section, 100 μm, 72-year-old man). Usually in these neurons two thick dendritic stems emerge from opposite lateral parts of the perikaryon; the long axis is mainly horizontally oriented. Side branches are sparsely spined (*insets* in *b* and *c*). The axon (*Ax*) arises at the basal portion of the cell body

the apical dendrite already ends near the soma and here gives rise to several side branches. In *d* a similar case occurs, the apical dendrite ascends obliquely and then terminates in several ramifications. In *c* the entire cell seems tilted to the left. The apical dendrite is slightly bent at the beginning and then takes an ascending course; beyond the first ramification its diameter is considerably reduced. In *e* the apical dendrite is only a stump from which two branches arise; the left one ascends. In examples *a*, *b*, and *d* the basal dendrites have a quite normal appearance, whereas in *c* and *e* they are directed upward

Large, Medium-sized, and Small Pyramidal Cells of Layer IIIc/IVa

The large (diameter up to 23 μm) and medium-sized (diameter 14–18 μm) pyramidal cells display "basophilic caps" and basophilic areas near the periphery of the soma, as is the case in the polygonal neurons. Since the long axis of their perikarya sometimes appears tilted to some extent, they cannot be unequivocally differentiated from the polygonal neurons. The cytoplasmic rim of the small pyramidal cells (diameter 9–13 μm) is faintly tinged.

Large Polygonal Neurons of Layers IIIc/IVa and IVb, Medium-sized Polygonal Neurons of Layers IIIc/IVa, IVb, and IVcα

The roughly triangular or polygonal perikaryon contains a spherical, smoothly contoured nucleus (diameter 11–13 μm or about 10 μm respectively; Fig. 20a–f) in a slightly eccentric position. Fine, evenly dispersed chromatin condensations give the nucleus a light appearance. Usually the nucleolus (diameter 3 μm or 2 μm, respectively) is attached to the nuclear membrane by a large chromatin clump. With the exception of some "basophilic caps" opposite the origins of the dendrites, the perinuclear zone of the cytoplasm is faintly tinged, whereas the peripheral zone is more intensely stained. The basophilic substance continues into the dendrites. The rarely encountered, large polygonal neurons correspond to the solitary Meynert-Cajal neurons.

Small Polygonal Neurons of Layers IIIc/IVa, IVb, and IVcα

The ovoid or spherical and smoothly contoured nucleus is surrounded by a narrow, faintly tinged cytoplasmic rim which contains only a few basophilic areas (Fig. 20g–k). All layers from IIIc to IVcα contain small polygonal neurons, in which the perikaryon is almost completely filled by the spherical nucleus and the faintly tinged cytoplasmic rim is extremely narrow (Fig. 20l–n).

Small, Heavily Pigmented Polygonal Neurons of Layer IVcβ

Around the light, spherical nucleus (diameter 7–9 μm) there is a very narrow and barely tinged cytoplasmic rim (Fig. 20o) which contains several typical lipofuscin granules (for details Braak H 1976, Braak E 1978).

8.3 Golgi Preparations

Within layer IV the following nerve cell types can be distinguished:
Py IIIc/IVa/1: large pyramidal cells
Py IIIc/IVa/2: medium-sized pyramidal cells
Py IIIc/IVa/3: small pyramidal cells
Poly-n IIIc/IVa-IVb/1: large polygonal neurons
Poly-n IIIc/IVa-IVb-IVcα/1: medium-sized polygonal neurons
Poly-n IIIc/IVa-IVb-IVcα/2: small polygonal neurons
Poly-n IIIc/IVa-IVb-IVcα-IVcβ: small polygonal neurons with delicate dendrites

Py IIIc/IVa/1 and Py IIIc/IVa/2: Large and Medium-sized
Pyramidal Cells of Layer IIIc/IVa

A stout apical dendrite emerges from the cell body, taking a bent or slightly twisted course and breaking up into its terminal tuft at the layer II/I boundary. Four to six side branches arise above and distally from the spine-free segment (about 50 μm) for a distance of about 70 μm. The dendrites are sparsely spined (Fig. 21a). The axon descends from a prominent axon hillock, gives off several collaterals, and projects to the white matter.

Py IIIc/IVa/3: Small Pyramidal Cells of Layer IIIc/IVa

The apical dendrite of these pyramidal cells extends only into layer IIIab. Here the impregnation ceases, and a terminal tuft often seems to be lacking. The apical and basal dendrites are sparsely endowed with stalked spines. Rarely neurons (diameter 11 μm) are impregnated, showing one basal dendrite opposite the apical dendrite (Fig. 23e). Both dendrites give off several side branches which are studded with stalked spines.

Poly-n IIIc/IVa-IVb/1: Large Polygonal Neurons Including the Meynert-Cajal Solitary Cells of Layers IIIc/IVa and IVb

From the polygonal cell body a number of stout dendrites radiate in all directions, with no preferential point of origin from the perikaryon (Fig. 21b—e). In the solitary Meynert-Cajal neurons (Fig. 22a—d) the lateral parts of the soma gradually transform into dendrites; they also bifurcate repeatedly in more distal parts of the parent dendrite. Most of their side branches remain within the layer and can be traced for about 300 μm; a few extend upward and downward for about 130 μm. The knobby dendrites are sparsely covered with stalked and stubby spines. The axon descends, giving off ascending collaterals.

Poly-n IIIc/IVa-IVb-IVcα/1 and Poly-n IIIc/IVa-IVb-IVcα/2: Medium-sized and Small Polygonal Neurons of Layers IIIc/IVa, IVb, and IVcα

In their overall appearance, the medium-sized (Fig. 23a—d) and most of the small (Fig. 24d, e) polygonal neurons seem to be merely diminutives of the large ones.

Poly-n IIIc/IVa-IVb-IVcα-IVcβ: Small Polygonal Neurons with Delicate Dendrites of Layers IIIc/IVa, IVb, IVcα, and IVcβ

Several small polygonal neurons occur with conspicuously delicate dendrites radiating in all directions, covered with long and short stalked spines, (Fig. 24a—c).

8.4 Electron Microscopy

The medium-sized and small pyramidal cells of layer IIIc/IVa do not differ from those of layer IIIab.

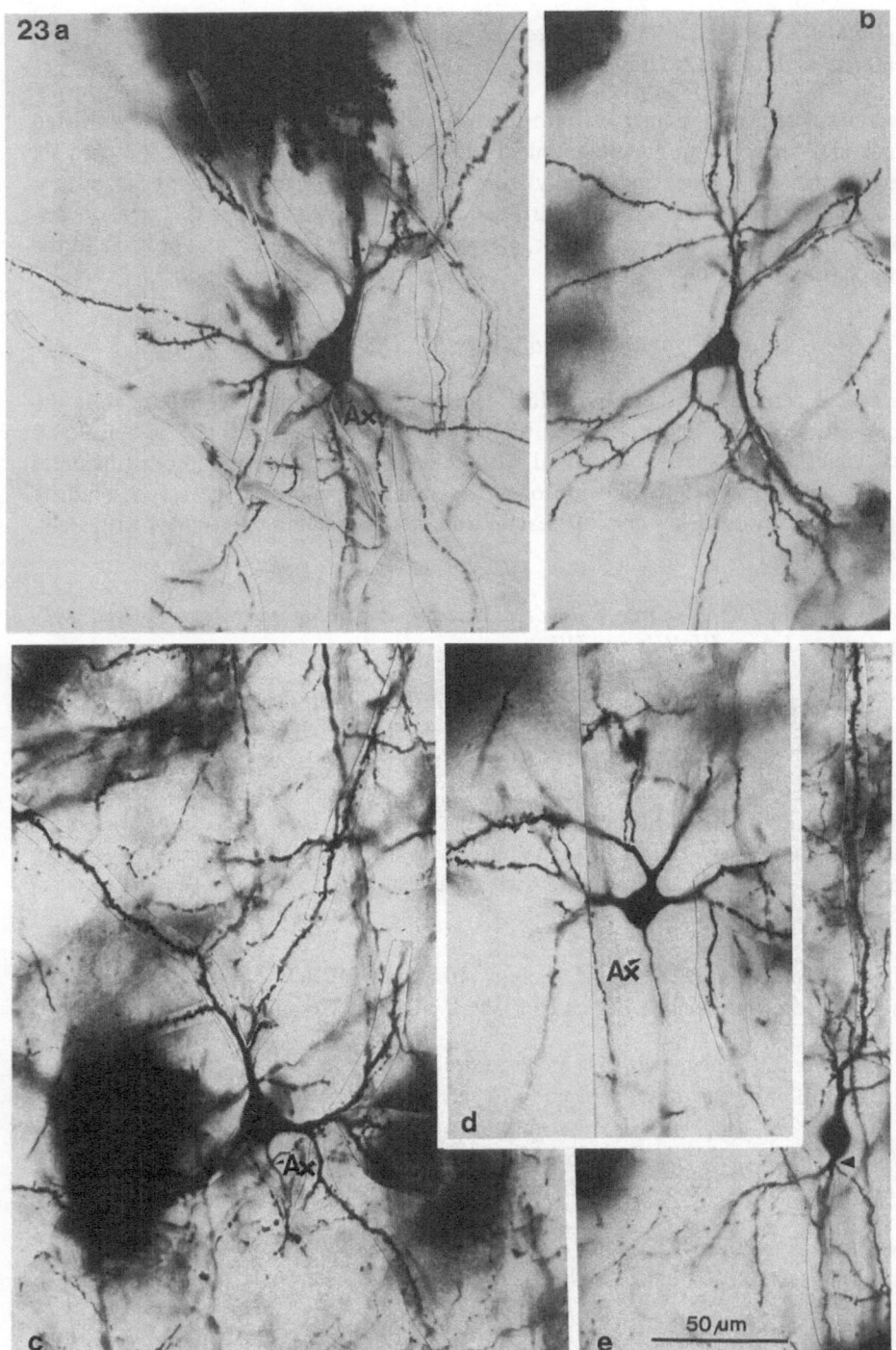

Fig. 23a–e. Photomicrographs of Golgi-impregnated medium-sized polygonal neurons (*a–d*: Poly-n IIIc/IVa-IVb-IVcα/1) and a small modified pyramidal cell (*e*: Py IIIc/IVa/3) of layer IIIc/IVa (frozen section, 100 μm; 72-year-old man). In the polygonal neurons an apical process of varying length gives rise to several side branches (*a–c*) or bifurcates close to the cell body (*d*). The basal

34

Large Pyramidal Cells of Layer IIIc/IVa and Polygonal Neurons of Layers IIIc/IVa, IVb, IVcα, and IVcβ

The large pyramidal and large polygonal neurons (Fig. 25), including the solitary Meynert-Cajal cells, show a perinuclear zone which contains extensive Golgi complexes and is almost devoid of rough endoplasmic reticulum and polyribosomes. In the peripheral zone of the soma, numerous cisterns of the rough endoplasmic reticulum are sometimes arranged in small stacks and in lamellar bodies (LeBeux 1972). Abundant polyribosomes, dense bodies, and mitochondria also occur. The mitochondria contain many cristae and have a medium-dense matrix. There are rare axosomatic synaptic contacts formed by boutons with flat vesicles. The axon hillock shows asymmetric and symmetric synaptic contacts, as well as subsurface cisterns, frequently associated with microtubules.

In sections through the middle of the perikarya of the medium-sized neurons (Fig. 26b), and especially of the small polygonal neurons (Fig. 26a), the nucleus occupies a proportionally large part of the cell body. The cytoplasm is confined to a rather thin rim with little space left for the evenly distributed organelles. The cisterns of the rough endoplasmic reticulum and the Golgi complexes are usually found in areas opposite the origin of the dendrites. The number of organelles extending into the dendrites decreases distally, while that of the microtubules increases. The medium-sized polygonal neurons are frequently associated with dark oligodendrocytes (Fig. 26b). Synaptic contacts at the soma membrane (Fig. 26a and inset), axon hillock, and initial segment are rarely encountered. For details concerning the small, heavily pigmented neurons of layer IVcβ see Braak E (1978).

With the aid of intracellular horseradish peroxidase injection, Gilbert and Wiesel (1979, cat) demonstrated that the overwhelming majority of the cells in layer IV were spiny stellate cells with an axon which gives off extensive arborization in layers II and III, and then projects beyond the cortex. In combined Golgi and electron-microscopical studies LeVay (1973) and Somogyi (1978, rat) have shown that boutons belonging to the axonal arborization of the spiny stellate cells establish asymmetric contacts, about 50% of which are found on smooth dendritic shafts of real nonpyramidal cells. Experiments using retrograde labeling have given some evidence that layer-IVb neurons project to area 18 (Rockland and Pandya 1979, *Macaca*) and to the middle temporal visual area (Spatz 1975, *Callithrix jacchus*).

In layer IVc of the macaque monkey striate cortex (Hubel and Wiesel 1972, 1977), where the bulk of the geniculo-cortical afferents terminate, all neurons belong to the least complicated cell types, the circularly symmetrical. Cells in layer IVc are almost exclusively monocular. Simple cells seem to be located mainly in layer IVb. The neurons of layer IVb are predominantly monocular.

---

dendritic field roughly encompasses a half sphere. All dendrites are moderately covered with stalked spines. The small pyramidal cell (*e*) is slightly modified, only one basal dendrite (*arrowhead*) arises from the cell body opposite the apical process. The apical dendrite bears numerous spines, the basal side branches are only sparsely spined. *Ax*, axon

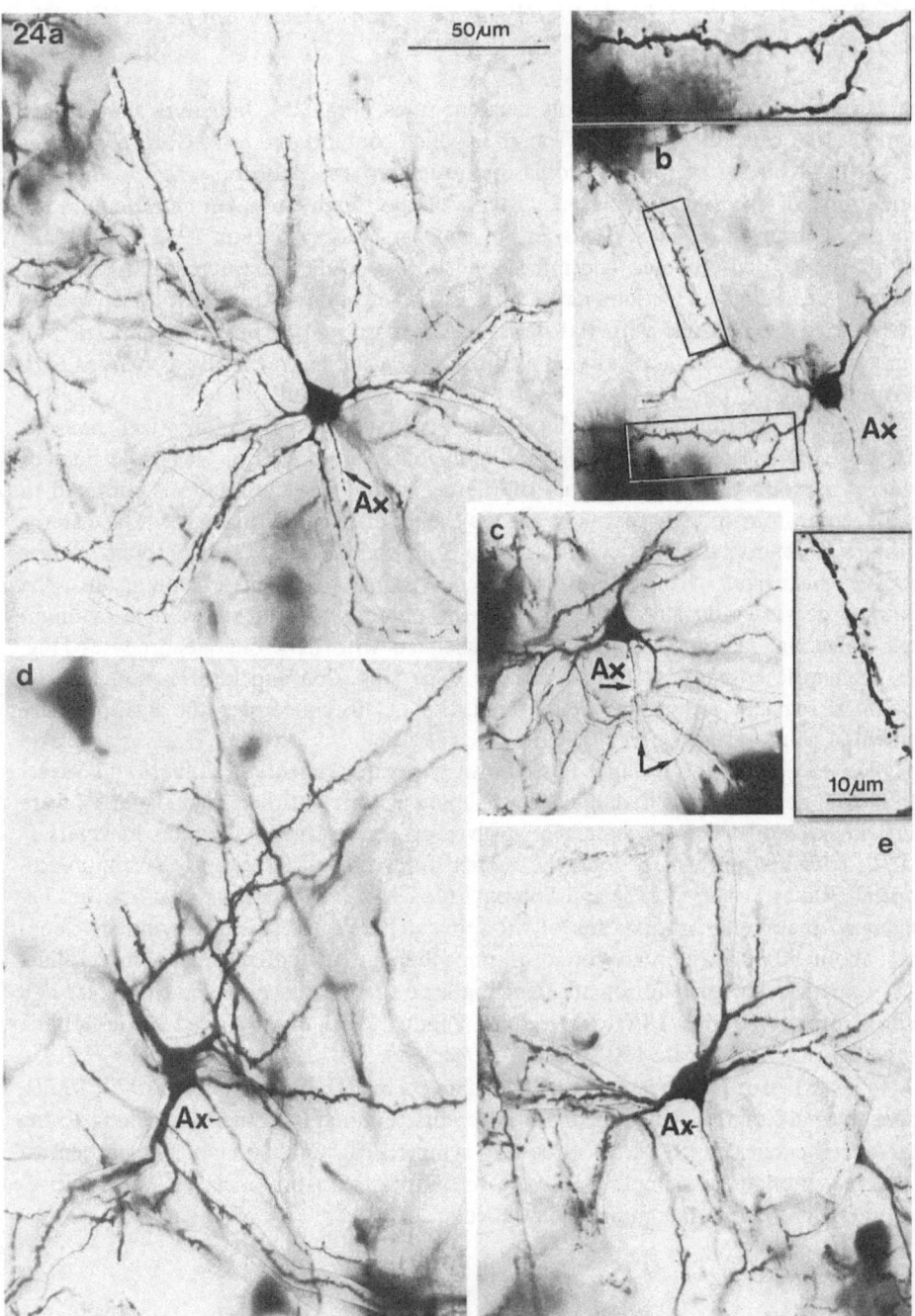

Fig. 24a–d. Photomicrographs of Golgi-impregnated small polygonal neurons (Poly-n IIIc/IVa-IVb-IVcα/2) of layers IIIc/IVa (d) and IVcα (e) and of small polygonal neurons with delicate dendrites (Poly-n IIIc/IVa-IVb-IVcα-IVcβ) of layers IIIc/IVa (a), IVb (b), and IVcα (c) frozen section, 100 μm; 72-year-old man). The dendrites bear few stalked spines. This is demonstrated by the *insets* at higher magnification from the *framed areas* in b. The axon (Ax) originates from the basal part of the cell body and gives rise to ascending collaterals (*arrows* in c)

36

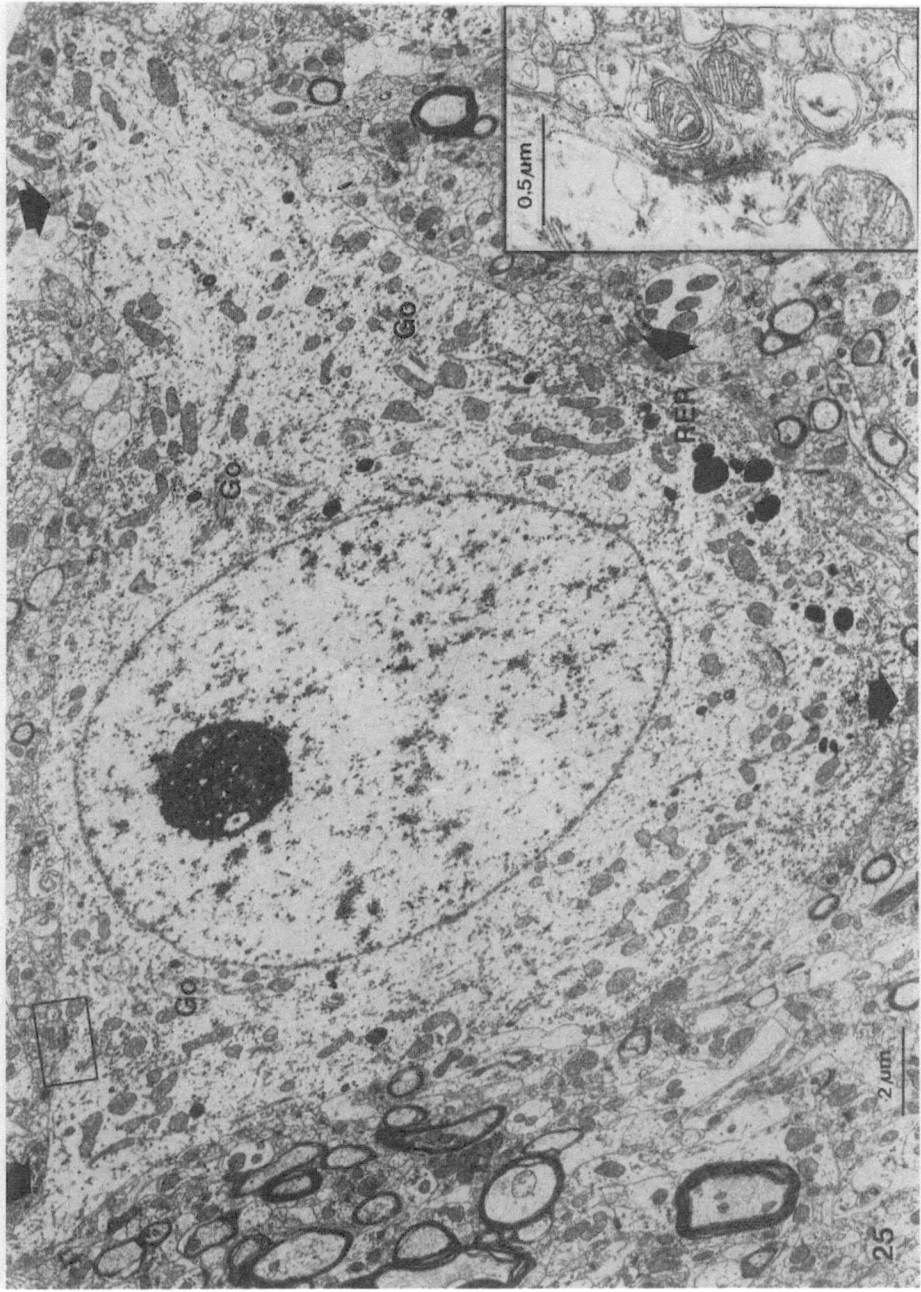

Fig. 25. Electron micrograph of a large polygonal Meynert-Cajal solitary neuron of layer IVb (66-year-old woman). RER cisterns (*RER*) are predominantly located in peripheral parts of the soma. *Go*, Golgi complexes. There are a few synaptic contacts on the soma surface and on the proximal dendritic segments (*arrows*). One of them (*framed area*) is shown at higher magnification in the *inset*

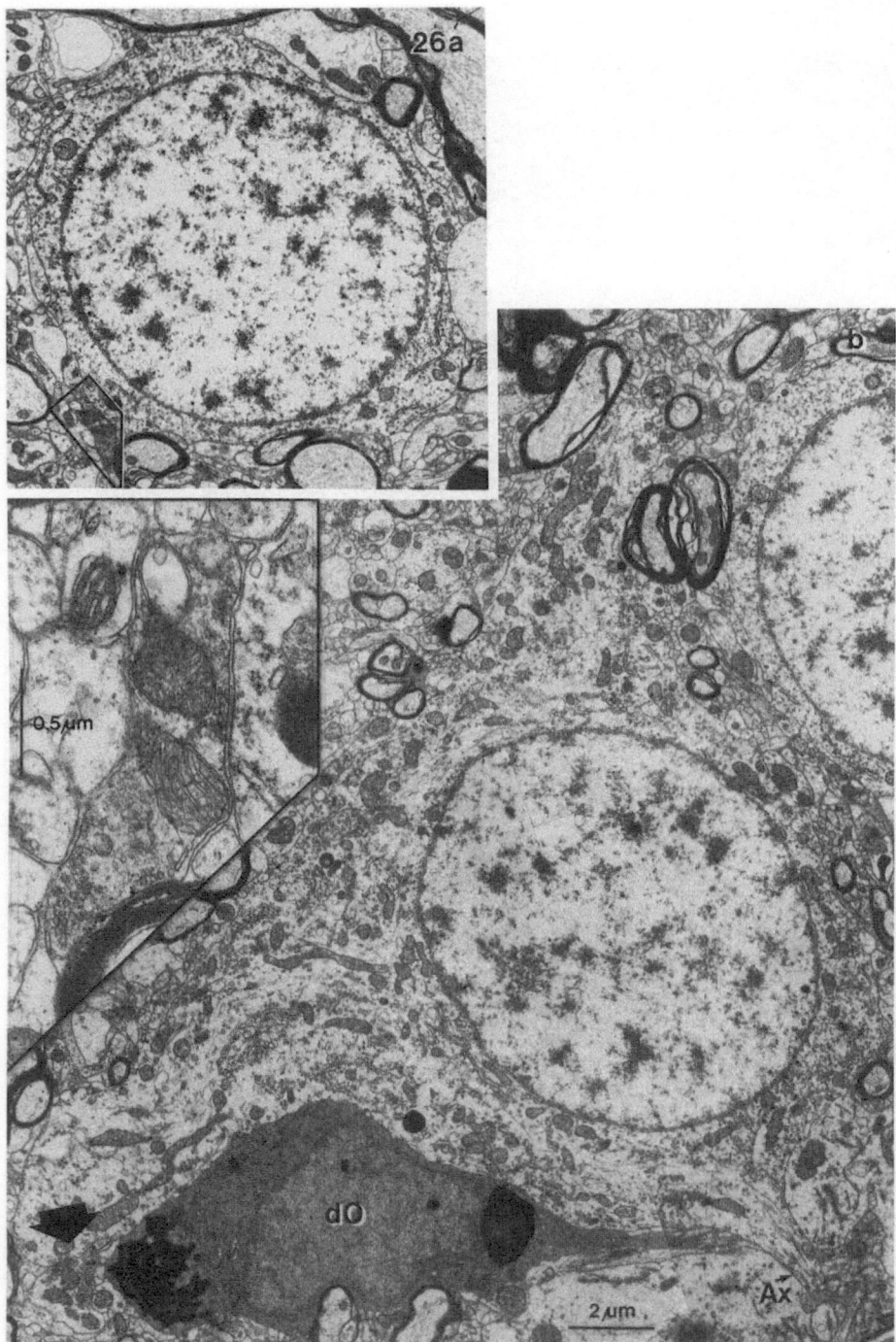

Fig. 26a, b. Electron micrographs of polygonal neurons (66-year-old woman). *a* Small polygonal neuron with a narrow cytoplasmic rim of layer IVcα. The framed area of a symmetric axo-soma-tic synaptic contact is demonstrated at higher magnification in the inset below. *b* Medium-sized polygonal neuron of layer IVb with one dendrite pointing to the *top* and another to the *bottom left*. At the tip of the latter, one symmetric synapse (*arrow*) is present. The axon (*Ax*) arises to the right of the basal part of the soma. A dark oligodendrocyte (*dO*) lies in close apposition to the soma and the dendrite

Numerous studies undertaken with different methods on various animals have demonstrated that the main afferent input to layer IV of the striate area originates in the corpus geniculatum laterale. To a considerably lesser extent, fibers coming from the claustrum (Carey et al. 1979, *Tupaia*) and from area 18 and other visual association areas, as well as from infragranular layers of the striate area itself, seem to terminate there. From degeneration studies following lesions of the corpus geniculatum laterale it is known that most fibers of the lateral geniculate nucleus (about 80%) terminate with asymmetric synaptic contacts on dendritic spines (Peters and Feldman 1976, rat; Peters et al. 1976a, rat; Schober and Winkelmann 1977, rat; Somogyi 1978: rat; Tigges and Tigges 1979, *Saimiri sciureus*), with fewer (about 15%) terminating on the dendritic shafts and very few (about 2%) on the somata. Most of the geniculate afferents synapse on spines emerging from thin dendrites (Peters and Feldman 1977, rat). These dendrites are basal side branches of layer III pyramidal neurons and apical side branches of layer V pyramidal cells or dendrites of spiny stellate cells or sparsely spined stellate cells (Peters et al. 1979). Surprisingly, only 10%—20% of the total number of synapses degenerate following lesions in the lateral geniculate nucleus (Garey and Powell 1971, monkey; Peters and Feldman 1976, rat; Tigges and Tigges 1979, *Saimiri sciureus*). The lateral geniculate body is not the only source of boutons with spherical vesicles and asymmetric synaptic contacts: the axon ramifications of the spiny stellate cells form similar boutons (LeVay 1973, *Macaca*; Somogyi 1978, rat).

In layer IIIc/IVa, afferents from the parvocellular layers of the corpus geniculatum laterale (Garey and Powell 1971, *Macaca*; Lund 1973, *Macaca*; Rezak and Benevento 1979, *Macaca*; Rowe et al. 1978, *Saimiri*) and from area 18 (Wong-Riley 1978, *Saimiri*) have been demonstrated.

Numerous horizontally arranged myelinated axons which form the stria of Gennari or the outer stripe of Baillarger characterize layer IVb. The axon plexus is derived from axon collaterals of supragranular neurons (Butler and Jane 1977, Clark and Sunderland 1939) and polygonal nerve cells of layers IVb and IVcα (Lund 1973). The length of these horizontal fibers can be up to 1 mm (Colonnier and Sas 1978, *Saimiri*; Fisken et al. 1975, *Macaca*). Relatively few synaptic terminals are visible in layer IVb. Retrograde labeling with horseradish peroxidase reveals a projection from the middle temporal visual area to layer IVb (Spatz 1977, *Callithrix jacchus*). Autoradiographic studies following injections into the infragranular layers also result in a labeling of layer IVb (Martinez-Millán and Holländer 1975).

In layer IVcα and IVcβ a large number of small dendritic profiles and dendritic spines are found. There are a few conspicuous, thick profiles (dendrites ?), sectioned longitudinally sometimes for about 10 μm, which receive numerous synaptic contacts (Fig. 8 and 9 in Braak E 1978). Among the numerous boutons in layer IVc occur some large ones (diameter 2—3 μm) which contain many mitochondria and evenly distributed spherical vesicles. The input from the lateral geniculate body terminates mainly in layer IVcβ, and to a lesser extent in layer IVcα (s.a. Spatz 1979, *Callithrix*). In detail, the magnocellular layers project to layer IVcα and the parvocellular layers to layer IVcβ (Glendenning et al. 1976, *Galago*; Hendrickson et al. 1978, old- and new-world primates; Hubel and Wiesel 1972, *Macaca*; Lund 1973, *Macaca*; Rowe et al. 1978, *Saimiri, Aotes*).

Ferster and LeVay (1978, cat) injected horseradish peroxidase into the optic radiation near the visual cortex. The enzyme diffused anterogradely and filled the axonal arborizations completely in area 17. Assuming that the axon diameter does not change on its way from the lateral geniculate body to the cortex, they concluded that axons with a large diameter of type-1 relay neurons of the lateral geniculate body (thought to correspond to Y-cells) preferentially arborize within layer IVcα, whereas the end ramification of the medium-sized axons of type-2 relay cells (thought to correspond to X cells) appear to be confined to layer IVcβ. Some of these axons seem to be myelinated almost up to their terminal ramification, since lesions in the corpus geniculatum laterale reveal degenerating myelinated fibers in horizontally oriented bundles within layer IVcβ (Tigges M et al. 1977, *Saimiri*; Tigges and Tigges 1979, *Saimiri*). Large boutons with spherical vesicles and asymmetric synaptic contacts are a characteristic feature of layer IVcβ (Tigges M et al. 1977, Braak E 1978, man). They contribute 13% of the total synapse population (Tigges M et al. 1977). They virtually disappear after lesions of the corpus geniculatum laterale. In such cases electron-lucent and electron-dense degenerating axon terminals, as well as large and small degenerating fibers, can be observed, which probably originate from different classes of lateral geniculate neurons (Tigges and Tigges 1979), i.e., from the Y and X cells which differ in morphological characteristics and physiological response properties.

8.5 Are the Polygonal Neurons Modified Pyramidal Neurons?

Real pyramidal cells display an apical dendrite and several basal dendrites, all invested with dendritic spines. The axon descends and gives off collaterals in a characteristic manner. Cortical nerve cells which show more or less marked variations from this typical appearance are often found. The apical dendrite in particular may be short and thin, and in some types of nerve cells can even be lacking. Globus and Scheibel (1967) found pyramidal cells in the deep layers of the rabbit's striate area which show apical dendrites in all degrees of reduction. The multipolar neuron of sector CA 4 of the cornu ammonis has neither apical nor basal dendrites. But it has several features in common with real pyramidal cells of sector CA 3, the most conspicuous one being densely packed microdendrites along circumscribed parts of the main dendrites. Therefore, the multipolar cells of sector CA 4 have been considered to be "modified pyramidal cells" (Lorente de Nó 1934, Ramón y Cajal 1909–1911). For other examples of modified pyramidal cells see H. Braak (1980) and H. Braak et al. (1976). As is shown in Figs. 21–24, the layer IV neurons of the striate area offer a wide range of transformational changes from real pyramids to neurons with dendrites radiating in all directions. Furthermore, as in pyramidal cells the dendrites dichotomize repeatedly in their distal parts. They also become thinner, proceeding from near the soma to the end.

In the early postnatal period the dendrites of all isocortical neurons are spined (Lund et al. 1977, *Macaca nemestrina*). During further development the number of dendritic spines is reduced most markedly in nonpyramidal cells and to a lesser degree in pyramidal cells and spiny stellate cells (Boothe et al. 1979, Lund et al. 1977). Postnatally, neither pyramidal cells nor spiny stellate cells show marked spine populations on the cell bodies or on the initial dendritic segments (Lund et al. 1977). In these areas only Gray's type-2 synapses have been demonstrated (LeVay 1973). The spines which appear during development may possibly indicate sites of type-1 contacts (Lund et al. 1977). In nonpyramidal cells which are devoid of spines in the adult (Jones 1975, LeVay 1973, Lund 1973, Meller et al. 1969), these initial spine contacts may be drawn down onto the dendritic shaft. The dendritic and somal surfaces of adult nonpyramidal cells are at least contacted by type-1 and type-2 synapses. Ramón y Cajal mainly investigated very young brains; possibly for this reason he saw no differences in spine density, and therefore did not use this characteristic feature to classify the neurons of layer IV.

Another important feature characterizing pyramidal cells is the course of their axon. Even the axon of inverted pyramids (Van der Loos 1965) initially ascends toward the pia and then returns, taking a descending course (Kirsche et al. 1973). The axon of polygonal neurons regularly arises from basal parts of the soma and also proceeds to the white matter, a feature which supports the assumption that polygonal nerve cells are closely related to pyramidal cells. Furthermore, the soma surfaces of both pyramidal and polygonal neurons are rarely contacted by synapses; only a very small number of Gray's type-2 contacts are present. Finally, the designation "spiny stellate cells" appears inappropriate. The polygonal nerve cells should be considered modified pyramidal cells.

9 Pyramidal and Polygonal Neurons and Neuropil of Layers IVd/Va and Vb

9.1 Nissl-Stained and Methylene Blue-Azure II-Stained Sections

There is a fairly sharp transition from layer IVcβ to layers IVd/Va and Vb due to a step-by-step reduction in the packing density of the cell bodies. Real pyramidal cells reappear, with particularly small ones predominating (Fig. 27). A few medium-sized and small polygonal neurons are still present. In addition to these types of nerve cells, the broad, cell-sparse layer Vb contains Meynert pyramidal cells, the largest neurons of the striate area.

Polygonal Neurons of Layers IVd/Va and Vb

The morphological characteristics of the polygonal neurons found in these layers correspond to those described in the foregoing layers.

Small-Pyramidal Cells of Layers IVd/Va and Vb

The small pyramidal cells tend to aggregate into clusters and to be associated with satellite glial cells (Fig. 28a–c). Within the faintly stained reticulate cytoplasm the ovoid nucleus (diameter 7–10 μm) is slightly eccentric. Especially at the apical part of the soma, several lipofuscin granules are present.

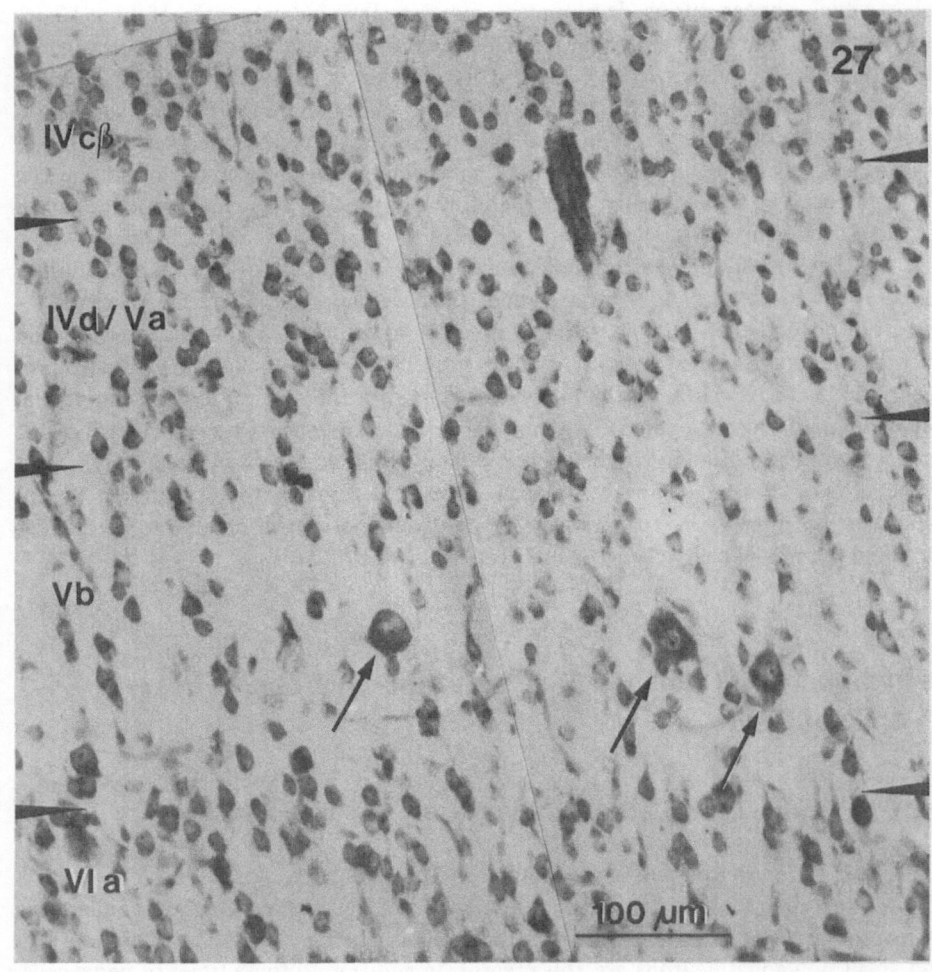

Fig. 27. Nissl-stained section of layer V of the striate area (Araldite, 10 μm, methylene blue; 66-year-old man). Layer IVd/Va is populated by small and medium-sized neurons, which are stained in varying intensity. In the cell-poor layer Vb the Meynert pyramidal cells (*arrows*) are prominent

Medium-sized Pyramidal Cells of Layer Vb

Within the medium-sized pyramidal cell bodies ($12-17 \times 22-30$ μm) the spherical or ovoid nuclei (diameter $11-13$ μm) are slightly eccentric. The cytoplasmic rim appears patchy, and intensely stained (Fig. 28d—f).

Fig. 28a—h. Nissl-stained pyramidal cells of layer IVd/Va and Vb (Araldite, 10 μm, methylene blue; 66-year-old man). Layer IVd/Va is predominantly populated by small pyramidal cells (*a—c*), which are often in clusters. Medium-sized pyramidal cells (*d—f*), as well as the large Meynert pyramidal cells (*g, h*), are present in layer Vb. *S,* satellite glial cell

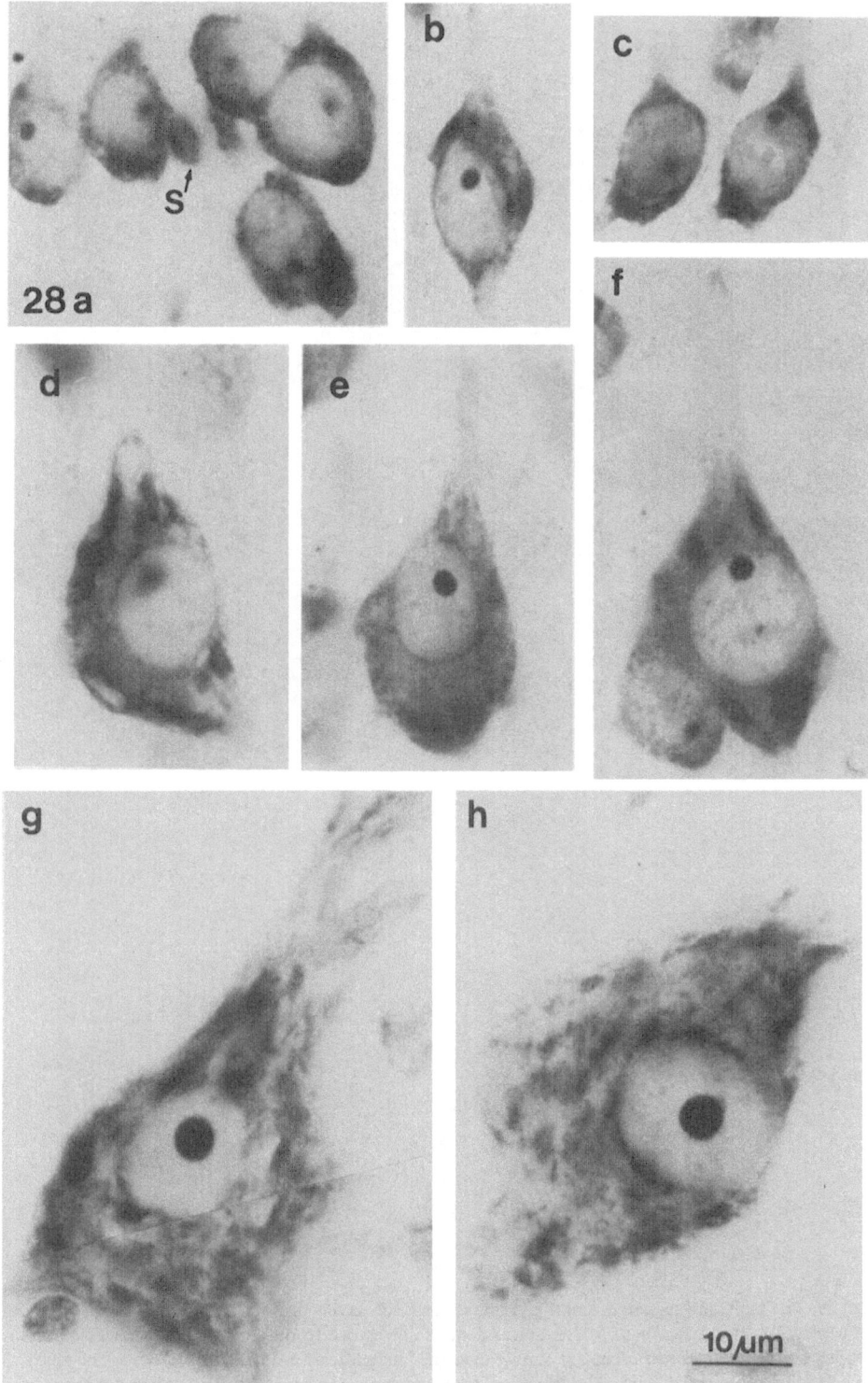

28 a

S↑

10 µm

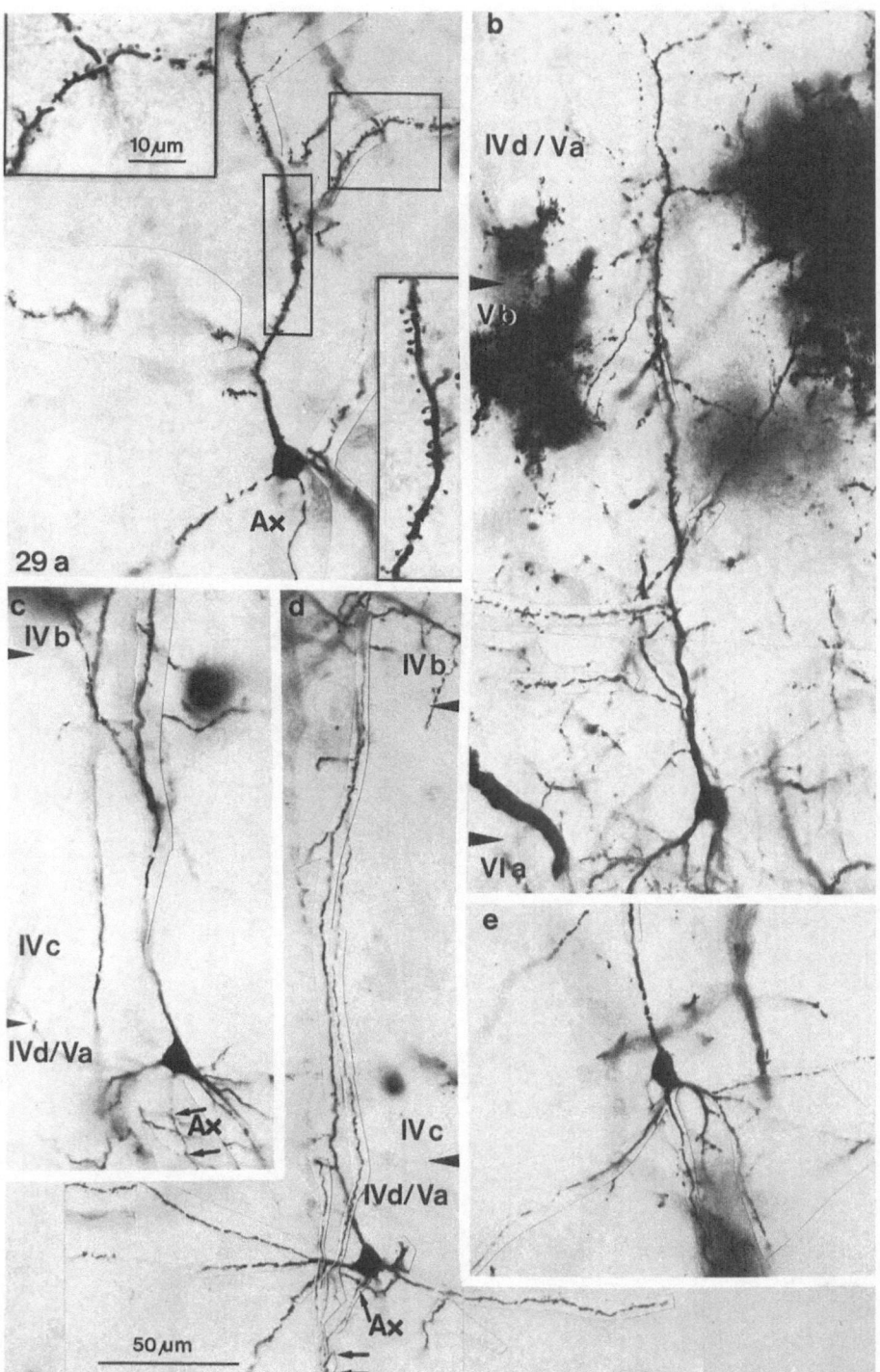

Fig. 29a–e. Golgi-impregnated small pyramidal cells of layer IVd/Va and Vb (frozen section, 100 μm; 72-year-old man). *a, b* The apical dendrite gives rise to only a few side branches; parts of both are spined (*framed areas* are shown at higher magnification in the *insets*). *c, e* The apical dendrites of these small pyramidal cells are thin and lack side branches. The apical dendrites as well as the basal dendrites are extremely sparsely spined. From the axon (*Ax*) ascending collaterals originate (*arrows*). The magnification bar of *d* is also applicable to *a–c* and *e*

The spherical nucleus (diameter 12−15 μm) of the Meynert pyramids may be localized in various parts of the large cell body (19−31 × 31−45 μm). It contains a prominent nucleolus (diameter 2.5−3.5 μm; Fig. 28g, h). Throughout the perikaryon, even adjacent to the nuclear membrane and in the proximal apical dendrite (for up to 30 μm), Nissl bodies of various size are frequently found. The lipofuscin granules accumulate at distinct areas of the soma. Often satellite glial cells are associated with the perikaryon and the apical dendrite.

9.2 Golgi Preparations

The following types of neurons are present:
Poly-n IVd/Va-Vb/1 and 2: medium sized and small polygonal neurons of layers IVd/Va
Py IVd/Va-Vb: small pyramidal cells of layers IVd/Va and Vb
Py Vb/1: medium-sized pyramidal cells of layer Vb
Py Vb/2: Meynert pyramidal cells

Poly-n IVd/Va-Vb/1 and 2: Medium-sized and Small Polygonal Neurons of Layers IVd/Va and Vb

The polygonal neurons found in these layers do not differ from those in the foregoing layer IV.

Py IVd/Va-Vb: Small Pyramidal Cells of Layers IVd/Va and Vb

In these layers there are at least two varieties of small pyramidal cells. One form (Fig. 29a, b) has an apical dendrite which gives rise to some side branches and appears to have some terminal twigs in layer IVa. All dendrites are sparsely covered with stalked spines. The descending axon is impregnated for a short distance only.

The other variety, lying predominantly in layer IVd/Va, has very delicate dendrites (Fig. 29c−e). The apical dendrite, which is often an unbranched process, appears to terminate at various levels of the layer IV. The basal dendrites are arrayed in a roughly hemispheric field (diameter about 200 μm). All dendrites are only sparsely covered with stalked spines. The descending axons give off ascending collaterals which extend to the fourth layer.

A type of small pyramidal cell which closely resembles these has been described by Lund and Boothe (1975, *Macaca mulatta*); these cells are likely to receive the axon projection of the spiny stellate neurons of layer IVcβ.

Py Vb/1: Medium-Sized Pyramidal Cells of Layer Vb

The medium-sized pyramidal cells of layer Vb issue a stout apical dendrite (Fig. 30a). It gives off very few, unbranched ramifications until impregnation ceases in layer IV; the basal dendrites are also sparsely ramified. The dendrites are moderately covered with long and short stalked spines. The axon descends from the base of the cell body.

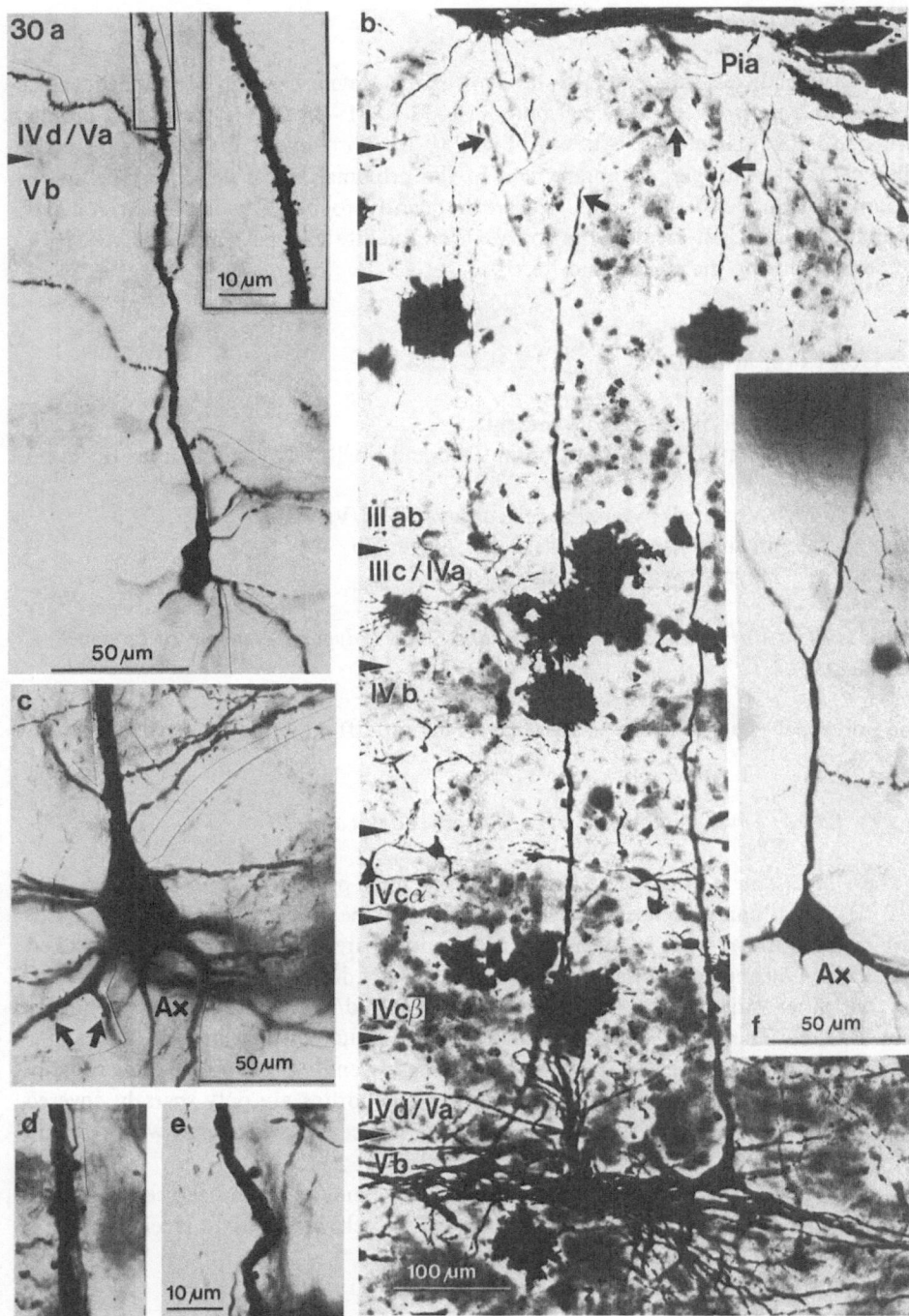

Fig. 30a–f. Photomicrographs of Golgi-impregnated nerve cells of layer Vb (frozen section, 100 μm; 72-year-old man, and in *b* only low viscosity embedding medium, 100 μm; 32-year-old woman). *a* The apical dendrite of the medium-sized pyramidal cells gives rise only to a few side branches, covered with stalked spines. *b* Low magnification of a group of Meynert pyramidal cells; their apical dendrites give rise to side branches only in the proximal part and bifurcate in the second layer (*arrows*). Most of the basal dendrites follow a horizontal course. *c* Meynert pyramidal cell

46

In the human adult, medium-sized pyramidal cells with densely spined dendrites and an apical dendrite ascending to the layer II/I boundary were not found as they were by Lund and Boothe (1975) in the macaque monkey. In autoradiographic studies layer IVb is labeled following injections which were made into the infragranular layers (Martinez-Millán and Holländer 1975, *Saimiri sciureus*). Intracellular injections of horseradish peroxidase (Gilbert and Wiesel 1979, cat) reveal a corresponding extensive axonal projection of layer V neurons to layer IV.

In addition, the medium-sized and small pyramidal cells of layer V project into parts of the pulvinar (Lund et al. 1975, *Macaca mulatta, M. nemestrina*; Ogren and Hendrickson 1976, 1977, 1979, *Macaca mulatta, Saimiri sciureus*; Raczkowski and Diamond 1978, *Galago senegalensis*; Trojanowsky and Jacobson 1977, rhesus monkey) and into the superior colliculi (Gilbert and Kelly 1975, cat; Lund et al. 1975, Magalhães-Castro et al. 1975, cat; Rhoades and Chalupa 1978, golden hamster).

Py Vb/2: Meynert Pyramidal Cells

The Meynert pyramidal cells are conspicuous due to their size. In general, the shape of their cell bodies (Fig. 30c) is not as uniform as it is in the large IIIab pyramidal cells. In some of the cases studied the apical dendrite extends to the layer II/I boundary where it bifurcates (Fig. 30b). In other cases the apical dendrite is delicate and begins to bifurcate in layer IVa (Fig. 30f). Human Meynert pyramidal cells do not seem to develop an "umbel of terminal dendrites" as has been shown by Chan-Palay et al. (1974) in *Macaca*. Only in layer Va does the apical dendrite give off side branches which themselves generally remain unbranched. From the lateral and basal parts of the cell body stout dendrites arise which extend for considerable distances (up to 600 μm) parallel to the surface, bifurcating repeatedly (Fig. 30b). The range of these basal dendrites considerably exceeds the distance between neighbouring perikarya (see also Chan-Palay et al. 1974). These basal dendrites form a conspicuous, tangentially oriented lattice (see also Ramón y Cajal 1900). However, the basal dendrites do not form bundles interconnected by membrane specializations as do the vertically oriented apical dendrites of the layer-V and VI pyramids (see Sect. 9.3). In human Meynert pyramidal cells a basal dendrite only occasionally runs obliquely downward for a short distance, whereas in monkeys, most of the basal dendrites descend (Chan-Palay et al. 1974, Lund and Boothe 1975). The dendrites bear only a very small number of stubby spines (Fig. 30c–e). In contrast to this, Meynert pyramidal cells of young humans (Conel 1939, newborn and 2-year-old child; Ramón y Cajal 1900, newborn) or animals (Chan-Palay et al. 1974, young adult *Macaca mulatta*) show heavily spined dendrites. But in 4- to 6-year-old children (Conel 1963, 1967) and in young macaque monkeys (*Macaca mulatta, M. nemestrina, M. fascicularis*: Lund and Boothe 1975) the number of dendritic spines is already considerably reduced. One is tempted to suggest that in this particular cell type the number of dendritic spines is more markedly reduced than in other pyramidal cells during postnatal development.

---

at the normal magnification, demonstrating that the basal dendrites also arise from the lateral parts of the cell body. The basal dendrites have only a few stubby spines (*arrows*). d, e Two segments of an apical dendrite of a Meynert pyramidal cell covered with a few stubby spines. f Unusual type of Meynert pyramidal cell with a poor dendritic ramification. The apical dendrite bifurcates in layer IVc

In most cases the impregnation of the axon ceases within layer V (Chan-Palay et al. 1974, *Macaca mulatta*; Clark 1942, *Macaca mulatta*; Conel 1939–1967, man; Lund and Boothe 1975, *Macaca mulatta, M. nemestrina, M. fascicularis*; Shkol'nik-Yarros 1971, man). Only Ramón y Cajal (1900) shows some axon collaterals returning from the white matter to layer V. Clark (1942) claimed that the axon of the Meynert pyramids project to the lamina tecti, whereas tracing methods with horseradish peroxidase gave much evidence that the axon or collaterals of the axon terminate in a cortical area buried within the superior temporal sulcus (Zeki 1971, 1976) thus forming the middle temporal visual area (Spatz 1975, 1977, *Callithrix jacchus*).

9.3. Electron Microscopy

Small and Medium-sized Pyramidal Cells

The smoothly surfaced nucleus is surrounded by a narrow cytoplasmic rim containing the usual organelles. Golgi complexes do not extend into the apical dendrite.

Meynert Pyramidal Cells

Within the broad cytoplasmic rim, large Golgi complexes form a shell around the smoothly contoured nucleus (Fig. 31). Small aggregates of cisterns of the rough endo-plasmic reticulum and polyribosomes are evenly distributed. Only the lipofuscin granules tend to lie in a cluster. Usually no more than two axosomatic synaptic contacts are present along the soma perimeter. The smoothly surfaced nucleus, the rare axosomatic synaptic contacts and the few neurofilaments in the proximal parts of the basal and apical dendrites in the human Meynert pyramids are not consistent with the findings of Chan-Palay et al. (1974) in Meynert pyramids of *Macaca mulatta*.

Neuropil

Characteristic features of layers IVd/Va and Vb are broad, radially oriented fascicles of myelinated fibers, as well as bundles of apical dendrites (Fig. 32a). Some of the large and small myelinated fibers are likely to be of extracortical origin, because they degenerate after lesions have been made in the lateral geniculate body (Tigges M et al. 1977, Tigges and Tigges 1979, *Saimiri*).

Often the radiate fascicles are accompanied by bundles of apical dendrites (Fig. 32a). These are the apical dendrites of layer-V and VI pyramidal cells which are in close apposition. Between adjacent dendritic membranes, puncta adhaerentia occur (Peters et al. 1976b, Smith and Moskowitz 1979; Fig. 32b). In addition, another type of membrane specialization can be found, showing a marked divergence of the facing dendrite membranes. A dense lamella runs through the interstitial cleft. The contrast of the membranes is enhanced, and at their cytoplasmic face they are accompanied by a light and dark line (Fig. 32b). Dendritic bundling presumably makes some form of interaction and synchronization possible within a certain group of neurons. They are possibly morphological equivalents of vertically arranged functional units which have been found in electrophysiological studies (Fleischhauer 1974, cat; Fleischhauer et al. 1972, rabbit; Peters and Walsh 1972, rat; Winkelmann et al. 1975, rat).

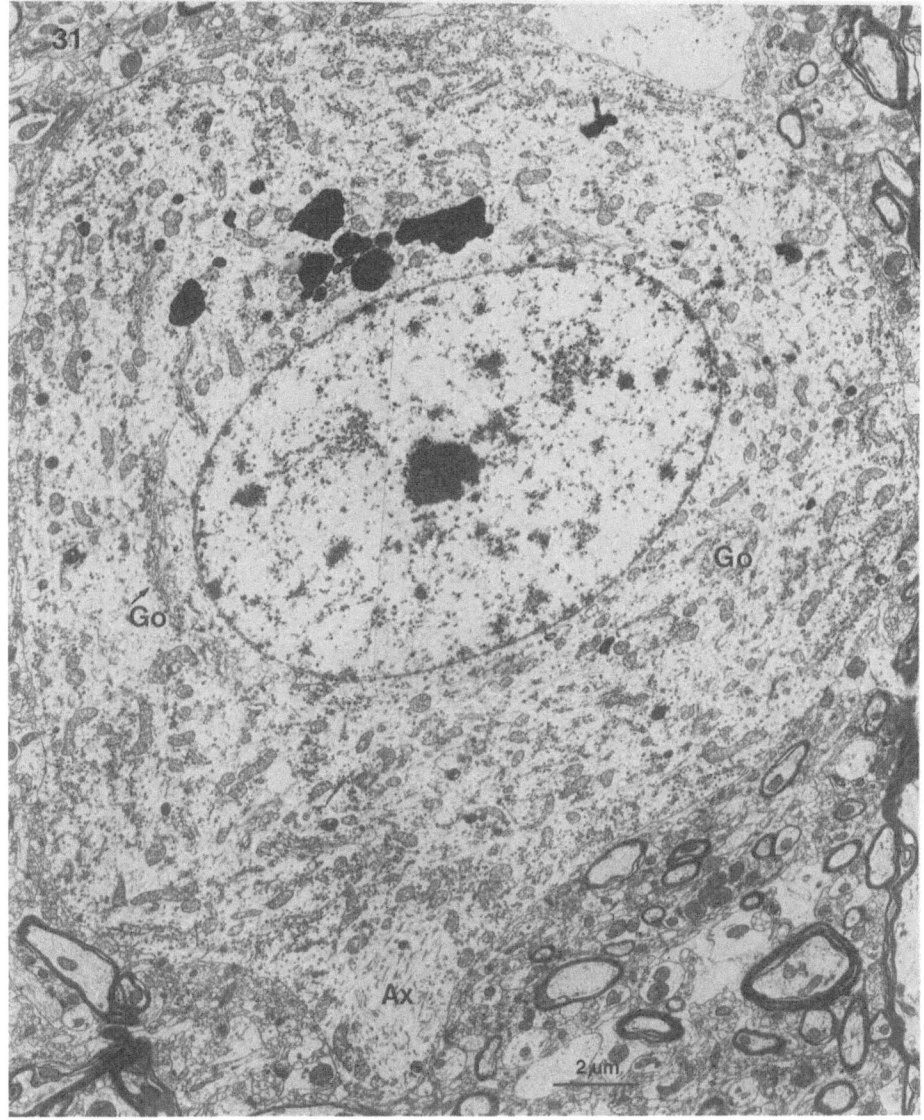

Fig. 31. Electron micrograph of a Meynert pyramidal cell (66-year-old woman) showing the axon hillock (*Ax*). *Go*, Golgi complexes

Boutons with pleomorphic vesicles have symmetric synaptic contacts on the apical dendrites. The postsynaptic site of these synapses is underlined with smooth, irregularly shaped cisterns. Within the neuropil, numerous boutons containing spherical vesicles form asymmetric synaptic contacts on small dendrites and dendritic spines.

The results of degeneration and autoradiographic studies lead to the conclusion that numerous collaterals of layers-II and III pyramidal cells terminate within layer V (Benevento and Ebner 1971, opossum; Butler and Jane 1977, rat; Butler et al. 1979, *Tupaia glis*; Martinez-Millán and Holländer 1975, *Saimiri*; Spatz et al. 1970; *Saimiri*).

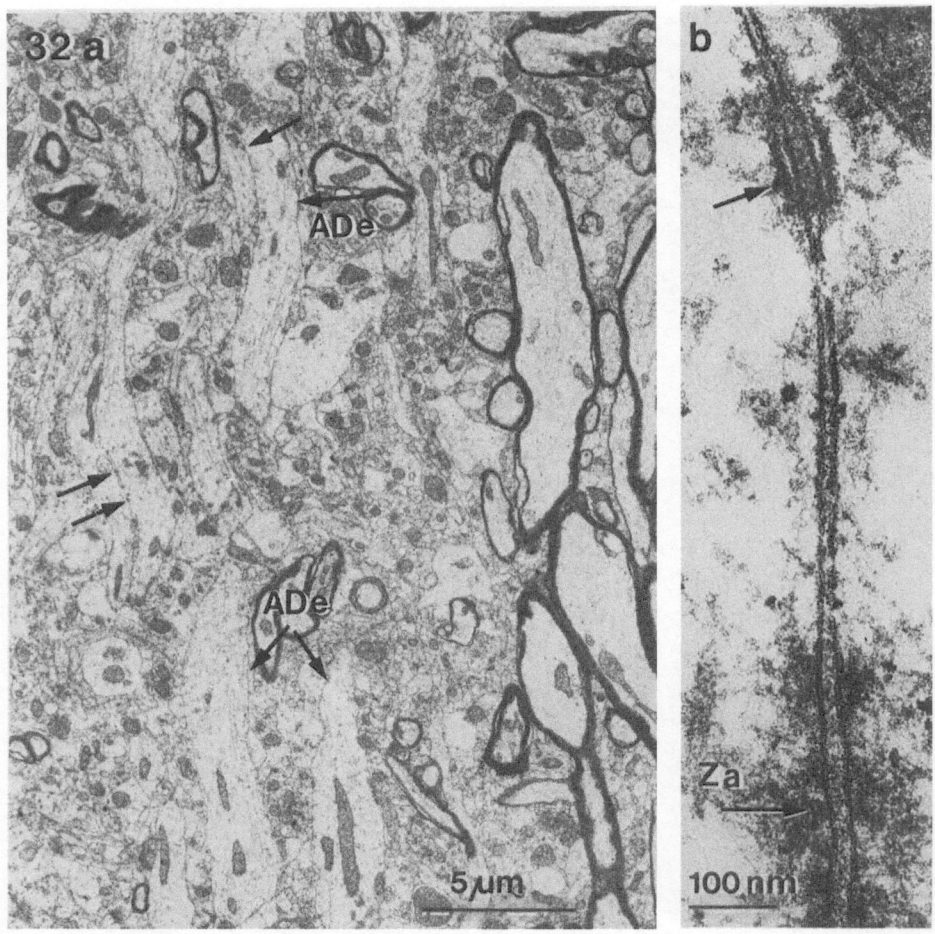

Fig. 32a, b. Electron micrograph of a radiate bundle of layer IVd/Va (66-year-old woman). *a* The apical dendrites (*ADe, left*) are connected by intercellular junctions (*arrows*). *b* High magnification of two facing membranes of two apical dendrites connected by puncta adherentia (*Za*) and another type of membrane specialization (*arrow*) in which the electron-dense bar is separated from the membrane by an electron-lucent line

Spines of small and medium-sized dendrites (about four-fifths) and dendritic shafts (about one-fifth) are postsynaptic to degenerating axon terminals following supra-granular lesions (Butler and Jane 1977, Butler et al. 1979). In addition, axon collate-rals of layer-IV neurons (Lund 1973, Lund and Boothe 1975) and afferents from area 18 (Tigges J et al. 1977, *Saimiri*; Wong-Riley 1978, *Saimiri*) terminate within the layer V.

10 Pyramidal Cells and Multiformed Neurons of Layers VIa and VIb

10.1 Nissl Preparations

Layer VIa is recognizable by its densely packed population of small and medium-sized nerve cells (Fig. 33). In layer VIb the number of cell bodies is slightly reduced, many of the perikarya appear larger than those of layer VIa, and some pyramids have a broad cytoplasmic rim. The perikarya look pyramid- or pear-shaped, or even fusiform. The long axis of most of the perikarya is tilted and the proximal apical dendrite gradually becomes vertical.

Small Pyramidal Cells of Layers VIa and VIb

In the small pyramidal cells (diameter 7−10 µm) the ovoid nucleus is surrounded by a faintly tinged cytoplasmic rim (Fig. 34a−c) which is narrow in some cases (Fig. 34g, right neuron). On rare occasions a nuclear fold indents the nucleus.

Medium-sized Pyramidal Cells of Layers VIa and VIb

Within the medium-sized (11−13 × 17−21 µm) perikarya, Nissl bodies and clusters of lipofuscin granules surround the nucleus, which is ovoid and sometimes indented by folds (Fig. 34d−g). Especially at the base of the apical dendrite a block of Nissl substance forms a "basophilic cap". Some of the medium-sized pyramidal cells display a broad cytoplasmic rim (16−24 × 21−26 µm) and basophilic patches near the soma periphery (Fig. 34h−i). In the Nissl preparations the group of medium-sized pyramidal cells fail to show any conspicuous differences which would permit distinction between the varieties of pyramidal cells seen in Golgi preparations.

10.2 Golgi Preparations

The following types of neurons are present:
  Py VI/1: small pyramidal cells
  Py VI/2: small triangular neurons
  Py VI/3: medium-sized pyramidal cells with a moderately spined apical dendrite
  Py VI/4: medium-sized pyramidal cells with an apical dendrite which is in part densely spined
  Py VI/5: medium-sized triangular neurons
  Py VI/6: medium-sized multiformed neurons

Py VI/1 and 2: Small Pyramidal Cells and Small Triangular Neurons of Layer VI

At least two varieties of small neurons are impregnated. In one form the apical dendrite ascends obliquely to layer Va (Fig. 35e). Apical side branches and basal dendrites are rarely found. All dendrites are sparsely spined (Fig. 35f). This type of small pyramidal cell was not mentioned by Lund and Boothe (1975) in their description of the monkey's striate area.

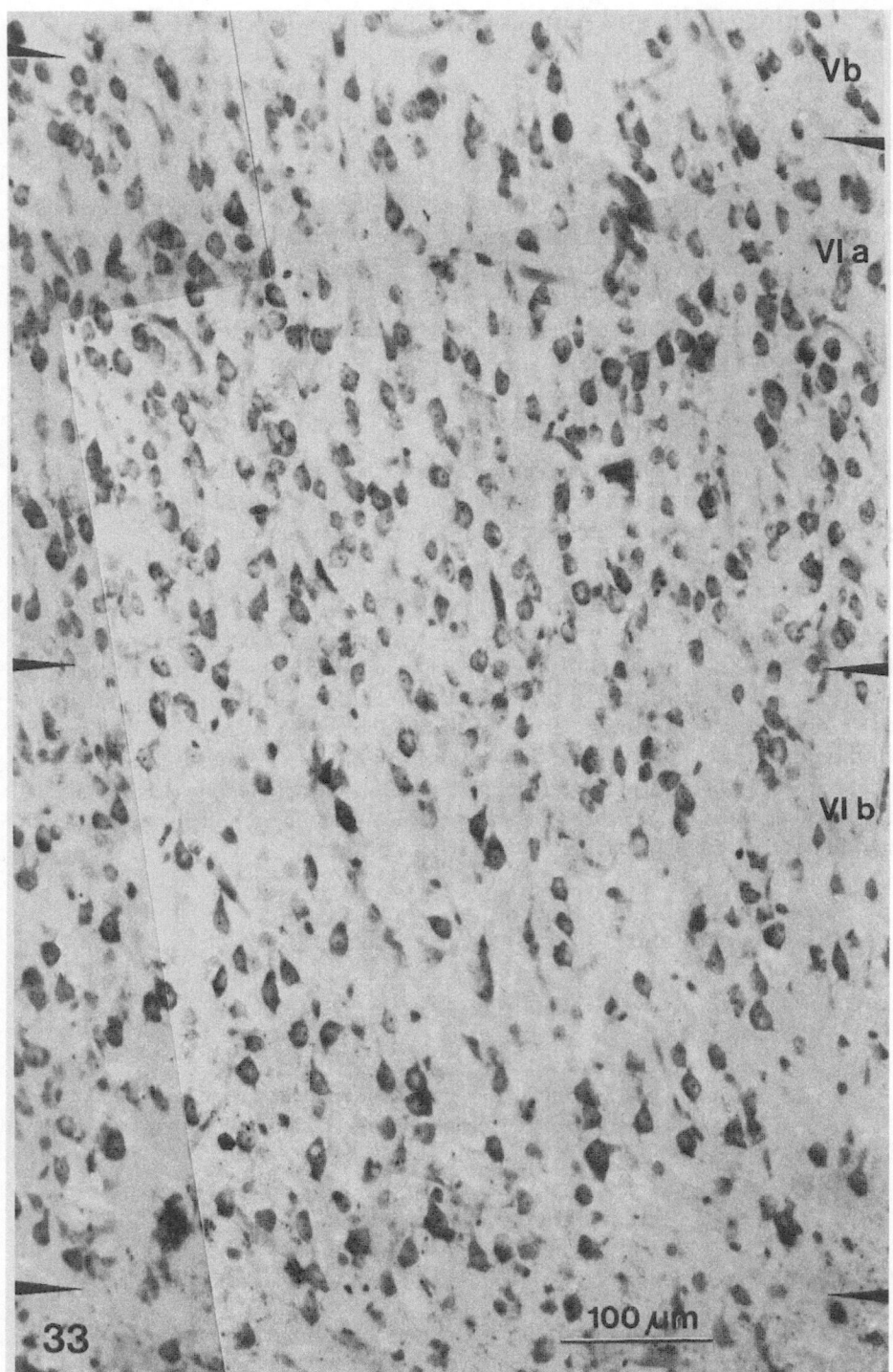

Fig. 33. Nissl-stained section of layer VI of the human striate area (Araldite, 10 μm, methylene blue; 66-year-old man). Layer VIa is densely populated by small and medium-sized pyramidal cells and multiformed neurons. In layer VIb the cell densitiy is reduced

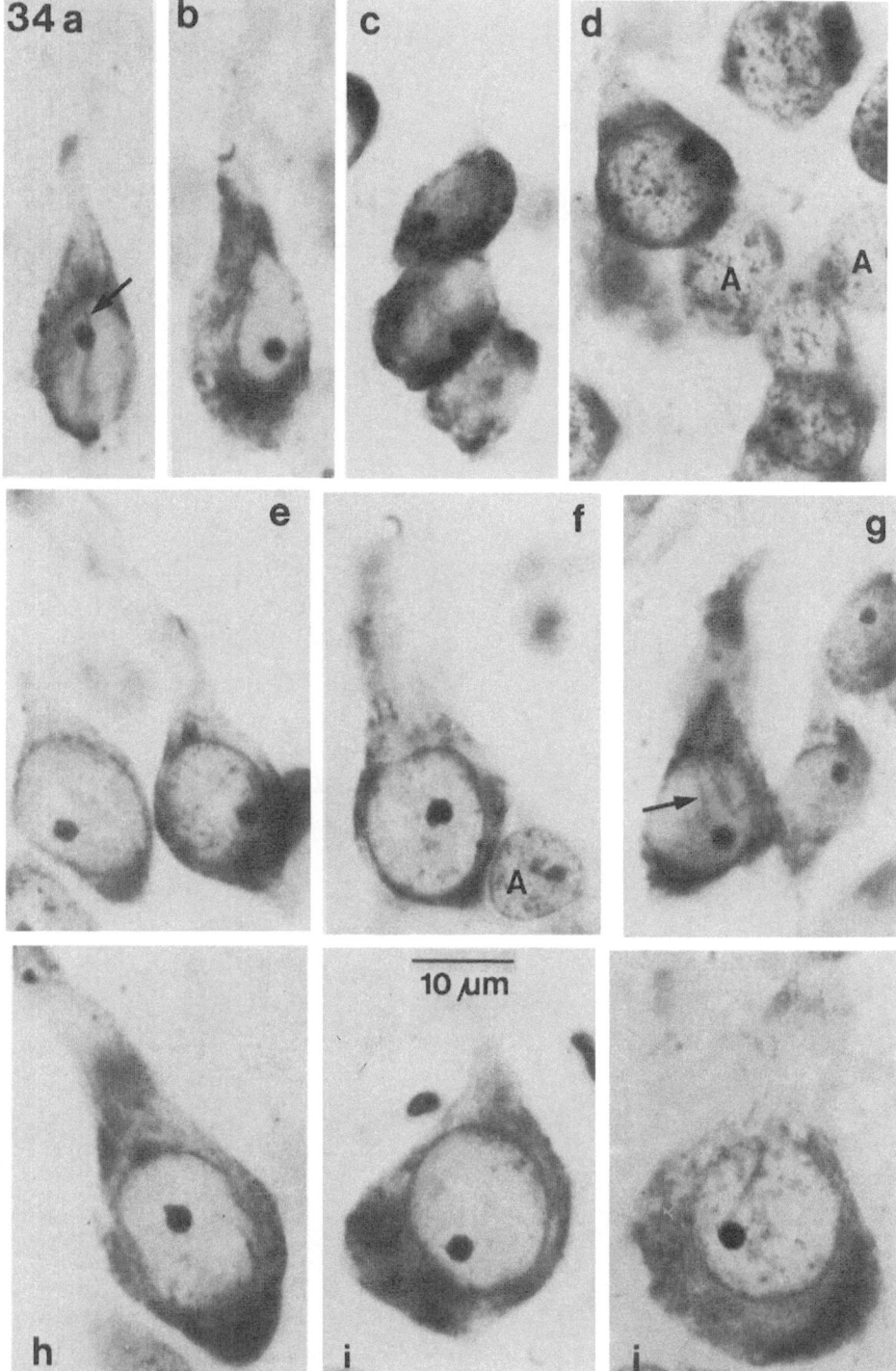

Fig. 34a–j. Nissl-stained nerve cells of layer VI (Araldite, 10 μm, methylene blue; 66-year-old man). *a–c* Small pyramidal cells. In *a* the nuclear fold is indicated by an *arrow*. *d–g* Medium-sized pyramidal cells, partly in close apposition to astrocytes (*A*). In *g* a nuclear fold is indicated by an arrow. *h–j* Medium-sized pyramidal cells with a broad cytoplasmic rim; basophilic material extends into the apical dendrite

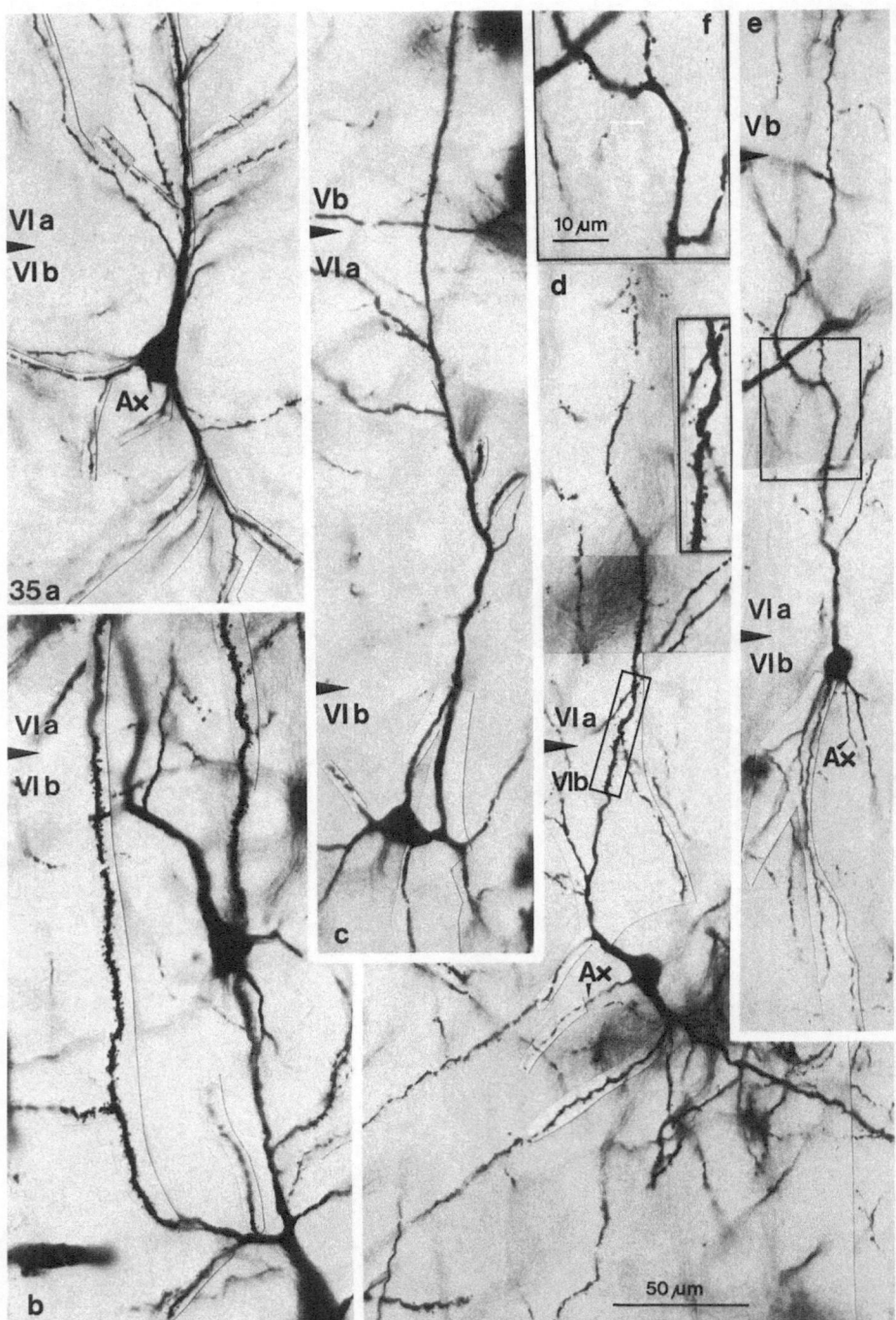

Fig. 35a–f. Golgi-impregnated nerve cells of layer VI (frozen section, 100 μm; 72-year-old man).
a Medium-sized triangular cell. *b* Modified, medium-sized pyramidal cell with an apical dendrite
bifurcating near the soma. Both side branches are covered with stalked spines. *c* Modified, medium-
sized pyramidal cell with the apical dendrite arising from the lateral part of the soma and having
its end ramification (not shown) in layer IVd/Va. *d* Small triangular cell with sparsely spined dendrites (see *framed area* at the higher magnification in the *inset* to the *right*). *e* Small pyramidal cell
with a sparsely spined apical dendrite, the *framed area* of which is shown at higher magnification in *f*

The other variety is the triangular cell (Ramón y Cajal 1900; Fig. 35d). This cell is characterized by two stout dendrites arising from the cell body almost opposite each other or at an obtuse angle; one of them takes an obliquely ascending, the other a descending course. These two main dendrites give rise to some scarcely ramifying side branches. One or two additional delicate dendrites arise from the cell body. All the dendrites are sparsely covered with stalked spines. The axon arises between the two main dendrites and descends.

Py VI/3, 4, 5, 6: Medium-sized Pyramidal Cells, Triangular and Multiformed Neurons of Layer VI

Four varieties of medium-sized nerve cells are found within layer VI. The most numerous are pyramidal cells; their apical dendrite is covered with a moderate number of spines. It gives rise to only a few side branches and shows terminal ramifications as early as in layer Va (Fig. 36a, c). One of the delicate side branches ascends up to layer IIIc/IVa. There are only a few sessile spines along the proximal apical dendrite (about 80–100 μm), whereas the more distal portions, as well as the apical side branches, show a moderate number of short stalked spines. Sparsely spined ramifying dendrites arise from the lateral and basal parts of the cell body.

A second type of medium-sized pyramidal cell shows apical dendrites which are particularly densely spined when traversing layer V (Fig. 36b, d, e). The proximal stem is sparsely studded with sessile spines, while the side branches arising in layer V are moderately covered with stalked spines. The apical dendrite becomes appreciably thinner in layer IVb without forming any terminal ramification. The basal dendrites resemble those of the first variety of cell.

Frequently, medium-sized triangular neurons (Fig. 35a) are impregnated. In general they resemble the small ones, but have numerous dendritic branchings.

The fourth type of medium-sized neuron shows some peculiarities in its dendritic tree and dendritic branching. In rare cases the cell body issues two ascending dendrites of equal thickness, both covered to a moderate degree with spines (Fig. 35b). They resemble the type-P_4 neurons of Lund and Boothe (1975). In another variety, one stout dendrite may also arise from the lateral surface of the soma, and then turn upward (Fig. 35c). Large fusiform neurons such as those described by Shkol'nik-Yarros (1971) have not been found in the material studied.

The first and second types of the human layer VI medium-sized pyramidal cells resemble the P_1 and P_2 neurons of Lund and Boothe, or the type-I and type-II neurons of Lund et al. (1977). In the opinion of Lund et al. (1977), the specific branching pattern of the apical dendrites (see also Schierhorn et al. 1973, rat) might be related to the incoming information from either parvocellular or magnocellular division of the dorsal lateral geniculate nucleus. Both types of pyramidal cells appear to collate information from different tiers of layer IV via their apical dendrites with information from layer VI via their basal dendrites.

The axon of layer-VI pyramidal cells with apical dendrites extensively branching in layer IV projects mainly into layer IV and terminates in the vicinity of the apical dendrite (Gilbert and Wiesel 1979, cat). This might partially explain the earlier observation of Martinez-Millán and Holländer (1975, *Saimiri sciureus*), who found layer IVb labeled after injections of horseradish peroxidase into infragranular layers.

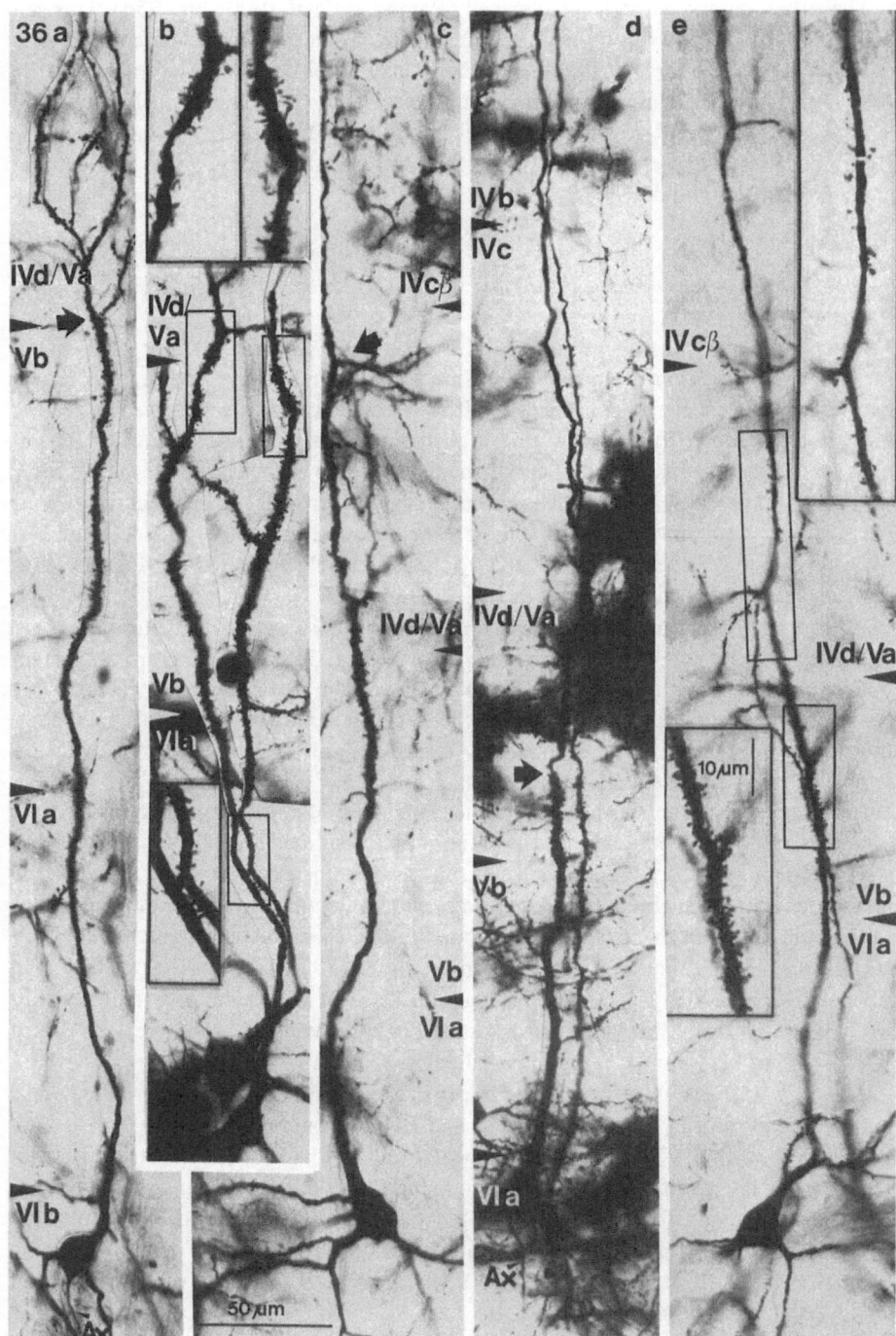

Fig. 36a–e. Photomicrographs of Golgi-impregnated medium-sized pyramidal cells of layer VI (frozen section, 100 μm; 72-year-old man). *a, c* The proximal part of the apical dendrite lacks side branches; in the upper part of layer Vb it is covered with spines and has ramifications in layer IVd/Va which extend into layer IVcβ. *b, d, e* Near the soma the apical dendrite gives rise to a few side branches and bears several spines (see *framed area* at higher magnification in the lower *inset* in *b*).

The pyramidal cells of the upper tiers of layer VI appear to provide major input into the parvocellular layers of the dorsal corpus geniculatum laterale, while the deeper parts project into the magnocellular layers (Gilbert and Kelly 1975, cat, Holländer 1974, *Saimiri sciureus*; Lin and Kaas 1977, *Aotes trivirgatus*; Lund and Boothe 1975, macaque monkeys; Lund et al. 1975, macaque monkeys; Raczkowski and Diamond 1978, *Galago senegalensis*; Robson and Hall 1975, squirrel monkey; Tömböl et al. 1975, cat; Trojanowsky and Jacobson 1977, rhesus monkey). Layer-VI neurons also seem to project into the claustrum (Carey et al. 1980, *Tupaia glis*).

Layer-VI neurons belong to the complex and hypercomplex cell types, over half of which receive input from both eyes (Hubel and Wiesel 1977). Layer VI sends a strong projection to the lateral geniculate nucleus, where each individual cell in each layer receives input from one eye only.

10.3 Electron Microscopy

Small Pyramidal Cells

A narrow cytoplasmic rim surrounds the nucleus, which sometimes has a single invagination (Fig. 37b). Hardly one axosomatic synapse occurs per section.

Medium-Sized Pyramidal Cells

Occasionally the nuclei have shallow indentations or one deep fold. Pyramidal cells with a broad cytoplasmic rim display up to three axosomatic synapses per section (Fig. 37a). The proximal apical dendrites are contacted by symmetric synapses at the shaft and by asymmetric ones at sessile spines (Fig. 37c).

Neuropil

The number of myelinated fibers bundled in radiate fascicles diminishes as layer VI is traversed from top to bottom, whereas the number of tangentially and obliquely oriented fibers increases. Some of the large and small radiating, myelinated fibers are of subcortical origin, since they degenerate after lesions which have been made in the lateral geniculate body (Tigges M et al. 1977, Tigges and Tigges 1979, *Saimiri sciureus*). Some of the tangentially oriented fibers are likely to have their origin in layers I to III. Perpendicular slits damaging the cortex from the surface down to layer III cause a moderate degree of fiber degeneration in layer VI (Fisken et al. 1975).

A few large synaptic boutons containing densely packed spherical vesicles with asymmetric synaptic contacts (type-2 boutons: Braak E 1978) and abundant small profiles with spherical vesicles and asymmetric contacts (type-4 boutons) are present.

The main afferents to layer VI come from the corpus geniculatum laterale (Gilbert

The apical dendrite is densely spined as it passes through layer Vb (see *upper framed areas* in *b*, at higher magnification in the *insets above* and in *d*, the *lower framed area* and the *lower inset* to the *left*). In layer IV the apical dendrite is sparsely spined (see *upper framed area* of *e* and *inset* to the *right*)

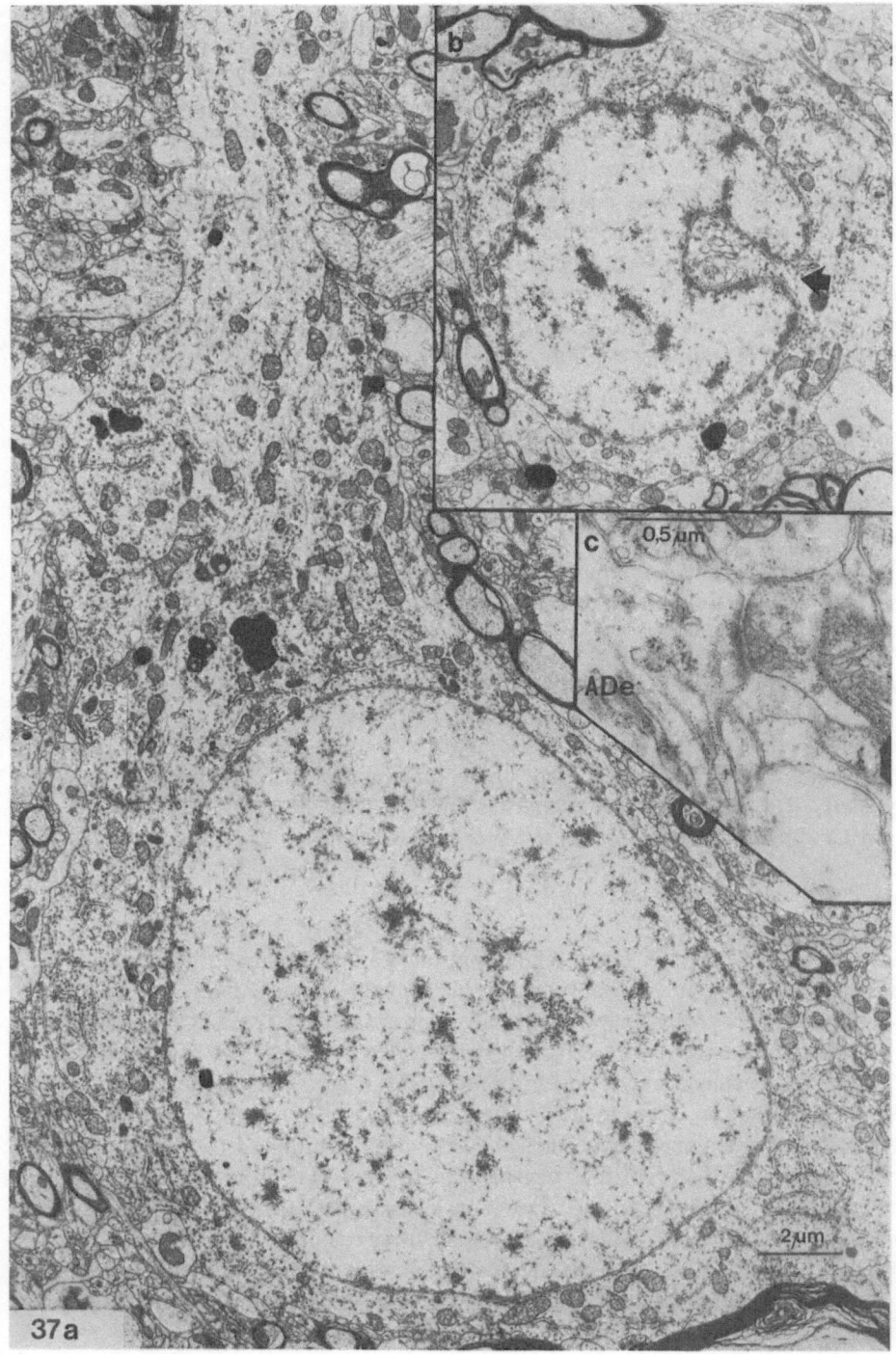

Fig. 37a—c. Electron micrographs of a medium-sized (*a*) and a small (*b*) pyramidal cell of layer VI (66-year-old woman). In *b* the *arrow* points to the nuclear fold. In *c* is a sessile spine of an apical dendrite (*ADe*) with an asymmetric synaptic contact

and Kelly 1975, cat; Lund 1973, *Macaca mulatta*; Peters and Feldman 1976, rat; Rezak and Benevento 1979, *Macaca mulatta*; Ribak and Peters 1975, rat; Rosenquist et al. 1974, cat; Tigges J et al. 1977, *Saimiri sciureus*). In all probability the geniculo-striate fibers make asymmetric synaptic contacts on dendritic spines, as has been shown in the rat by Peters and Saldanha (1976) and in the cat by Winfield and Powell (1976). Fibers from the pulvinar (Benevento et al. 1975, *Macaca mulatta*) and from the claustrum (Carey et al. 1980, Tupaia glis), as well as from area 18 (Tigges J et al. 1977, *Saimiri sciureus*; Wong-Riley 1978, *Saimiri sciureus*) and from the middle temporal visual area (Spatz 1977, *Callithrix jacchus*) terminate in layer VI.

11 Nonpyramidal Cells of Layers II to VI

11.1 Golgi Preparations

Compared with the perikarya of the pyramidal cells, those of the nonpyramidal cells are markedly heteromorphic (see also Table 3). The dendrites emerge from scattered points all over the surface and radiate in various directions. The dendrites arise in three different ways from the cell body. Thick dendritic stems, which gradually taper as they extend distally, arborize repeatedly at short distances from one another near the soma (Fig. 38c, d, e). Less frequently, on two opposite poles the perikaryon turns into two tapering processes, which give it a fusiform appearance (Fig. 40 d, f, g, h). Other thick dendrites emerge abruptly from the cell body; they also bifurcate repeatedly close to the parent soma (Fig. 38a, b, f). A third type of dendritic process is delicate and usually remains unbranched (Fig. 38a, b).

The branching of the dendrites is never profuse. Ramifications beyond a distance of 20 μm are rare (see also Sholl 1956). The smoothly contoured dendrites have dilations and constrictions at irregular intervals giving them a somewhat varicose appearance. The dendrites show a relatively uniform thickness which is maintained up to the most distal portions. The dendrites of most types of adult nonpyramidal cells are almost devoid of spines. On closer examination, a few of the dendrites may have single appendages close to the soma, but since these are thin and lack a terminal knob, it is questionable whether they are real spines.

The axon arises at any point on the perikaryal surface. Most often its impregnation already ceases at a distance of 10–20 μm from the soma. Only a few nonpyramidal cells have an axon which extends for about 30 μm, arborizing in a profuse plexus most often confined to the limits of the territory occupied by the dendrites.

Classification of nonpyramidal cells is based on the shape, size, and orientation of the cell body; the number, diameter, and course of the dendrites; and the arborization of the axon. Jones (1975, somato-sensory cortex of adult *Saimiri*) described six types of nonpyramidal cells with smooth-surfaced and sparsely spined dendrites which could be distinguished from each other by the characteristic arborization of the axon. A similar attempt at classification of neurons with intracortical axons has been done by Valverde (1976, mouse, visual cortex). Unfortunately, in the adult human cerebral cortex the impregnation of the axon often ceases in the vicinity of the soma and the other characteristics mentioned above must therefore be used for classi-

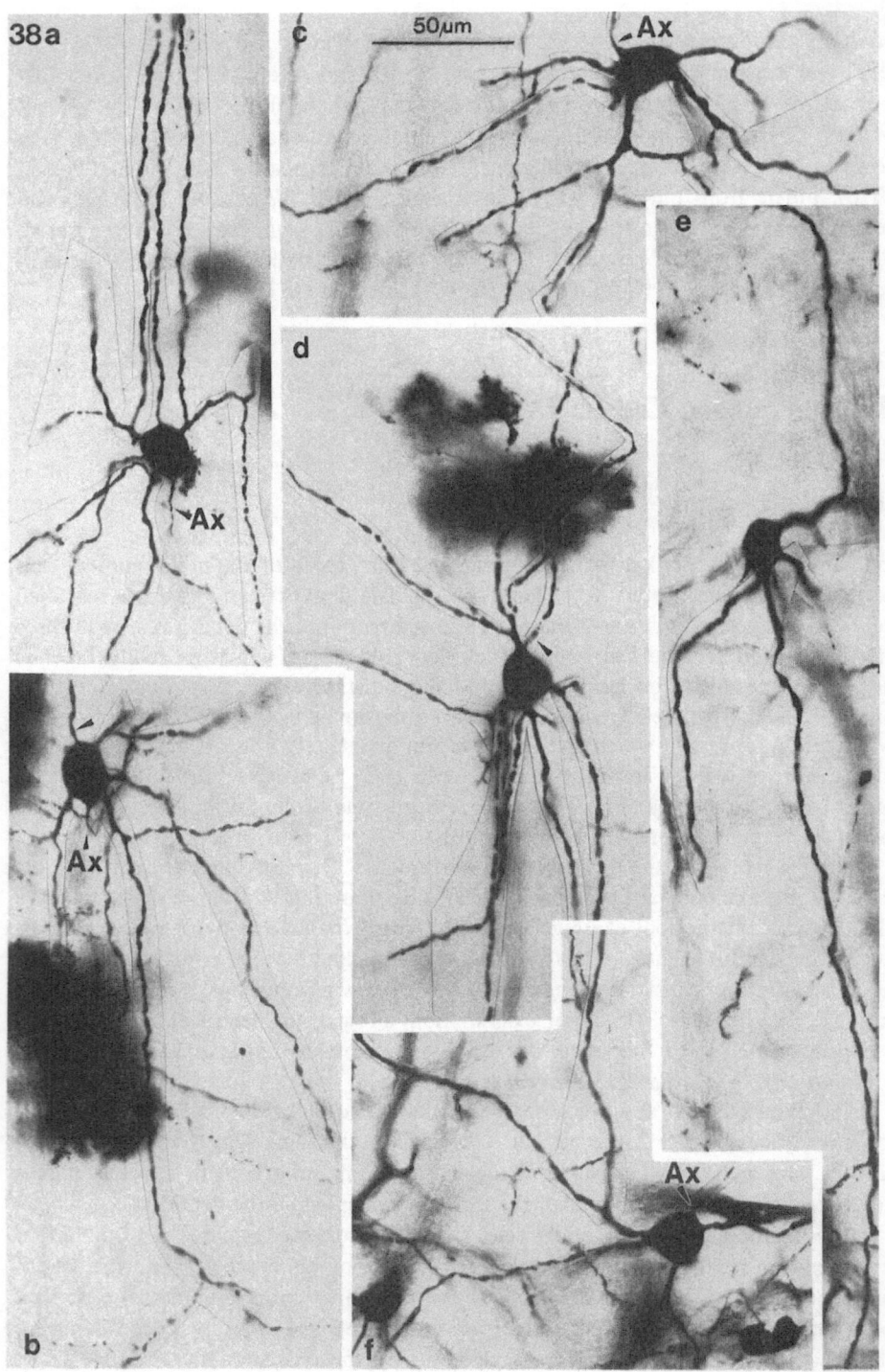

Fig. 38a–f. Photomicrographs of Golgi-impregnated large nonpyramidal cells of layers II (*a*), IVcβ (*b*), IVb (*c*), IIIab (*d*), V (*e*), and VI (*f*) (frozen section, 100 μm; 72-year-old man). Thick dendritic stems, which bifurcate repeatedly near the cell body, and thin nonramifying dendrites (*arrowheads*) arise from the soma. *Ax*, axon

fication. The most numerous nonpyramidal cells are large (Nonpy/1; Fig. 38a–f) and small (Nonpy/2; Fig. 39a–g) ones with rounded, ovoid, or irregularly shaped perikarya, with sizes given in Table 5.

The greatest diameter of large nonpyramidal neurons is roughly 30 μm. This is about half the size of the stellate cells recorded by Peters (1971) in the parietal cortex of the rat. The dendrites of the large and small nonpyramidal cells radiate in various directions. Less frequently, nonpyramidal cells occur, whose longer axis is oriented perpendicularly (Nonpy/3; Fig. 40d–h) or parallel (Nonpy/4; Fig. 40a–c) to the cortical surface; their dendrites usually arise from the two elongated poles of the cell body. The Nonpy/3-type neurons correspond roughly to the "cellules fusiformes à double bouquet dendritique" of Ramón y Cajal (1909–1911), and some of their features resemble type 3 of Jones (1975) and cell 3 of Peters and Fairén (1978). Their dendrites can usually be traced for about 250 μm, and pass through several layers. This type of nonpyramidal cell is possibly significant for connections within the cortical columns (Colonnier 1966, Jones 1975, Lorente de Nó 1938, Ramón y Cajal 1909–1911, Szentágothai 1971). The Nonpy/4-type cell in Fig. 40a resembles the stellate cell depicted by Lund (1973, *Macaca mulatta*, Fig. 33), which also has the same localization. Only on rare occasions can small nonpyramidal cells with numerous fine, short, ramifying dendrites (Nonpy/5) be found, radiating in all directions (Fig. 39h). These neurons resemble the neuroglioform cells of Ramón y Cajal (1909–1911). Another rarely occurring type has dendrites which curve around the perikaryon (Nonpy/6; Fig. 39i), similar to a type of stellate cell described by Lund (1973 in Fig. 17).

11.2 Nissl Preparations

The perikaryon of nonpyramidal cells is rounded or heteromorphic, and is generally more intensely stained than that of the pyramidal cells (Figs. 41a–m and 42a–s). The basophilic material is evenly distributed but does not extend into the dendrites (Figs. 41, 42; marked by asterisks). The eccentrically located nucleus displays shallow to deep folds of its membrane. Nissl bodies tend to lie within these indentations (Figs. 41, 42; marked by arrows). The entire nucleus of the nonpyramidal cell is more intensely stained than that of the pyramidal cell. Several small chromatin condensations adhere to the nuclear membrane, thereby emphasizing it. The ratio between the size of the nucleus and the width of the cytoplasmic rim is not as easily determined as it is in the

Table 5. Sizes of large and small nonpyramidal cells: the largest diameter and the diameter perpendicular to it

Layer	Large cells (Nonpy/1)		Small cells (Nonpy/2)	
II	16–22 × 14–18 μm	(n = 9)	14–19 × 11–14 μm	(n = 7)
IIIab	17–24 × 11–18 μm	(n = 15)	13–17 × 9–14 μm	(n = 16)
IIIc/IVa	17–28 × 16–21 μm	(n = 10)	14–17 × 13–15 μm	(n = 6)
IVb	17–30 × 13–20 μm	(n = 16)	13–16 × 11–14 μm	(n = 4)
IVc	19–21 × 14–16 μm	(n = 2)	14–19 × 11–14 μm	(n = 6)
V	16–29 × 11–22 μm	(n = 17)	13–16 × 11–14 μm	(n = 6)
VI	19–30 × 15–20 μm	(n = 8)	10–17 × 9–12 μm	(n = 2)

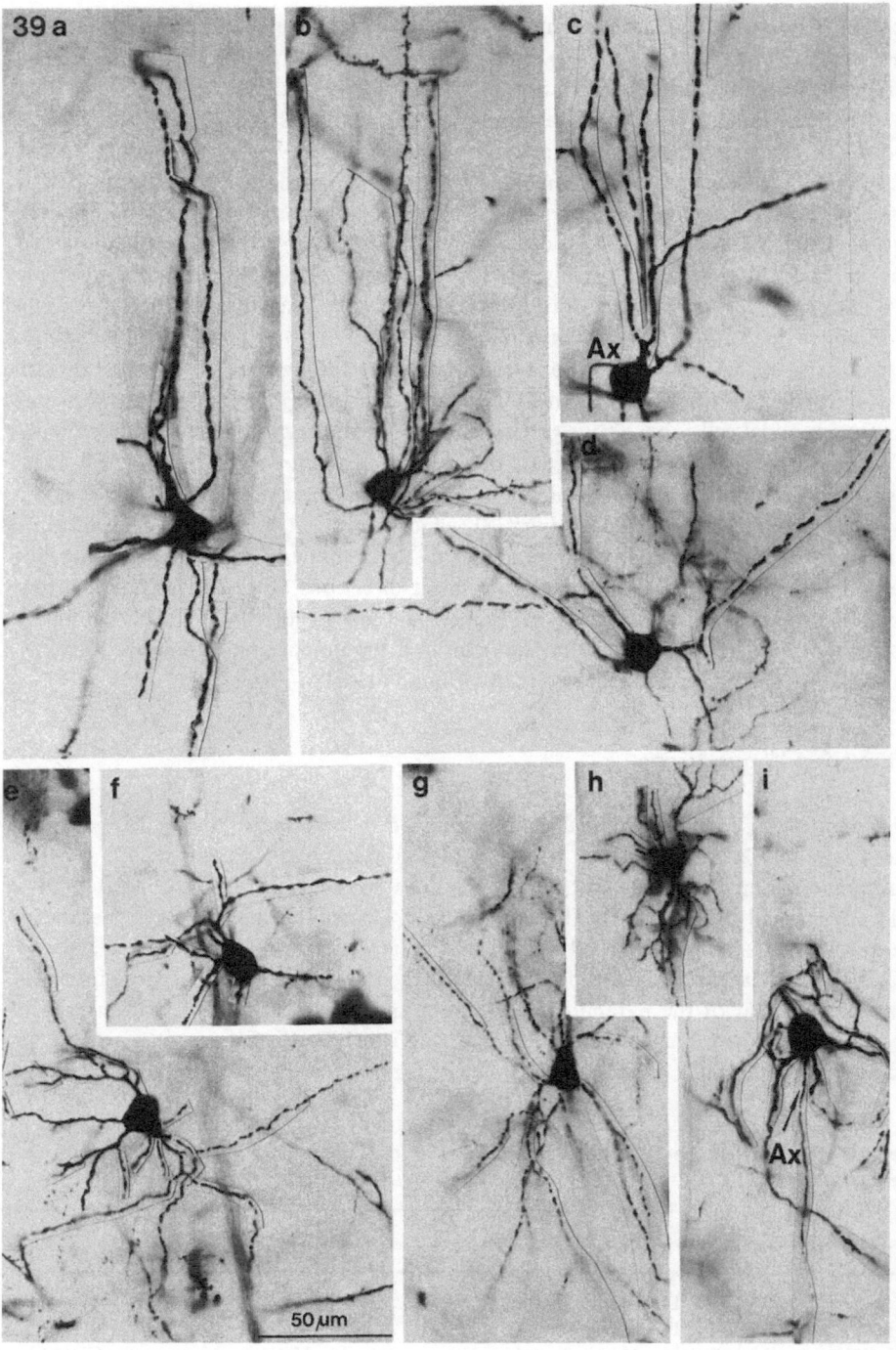

Fig. 39a–i. Photomicrographs of Golgi-impregnated small nonpyramidal (66-year-old woman) of layers II (*b*), IIIab (*a, c*), IIIc/IVa (*f*), and IVcα (*d*). Some of the small nonpyramidal cells have several delicate dendrites; the examples are from layers IVb (*e*), IVcβ (*g*), and II (*h*). One example of a small nonpyramidal cell with dendrites curving around the soma is shown in *i*. *Ax*, axon

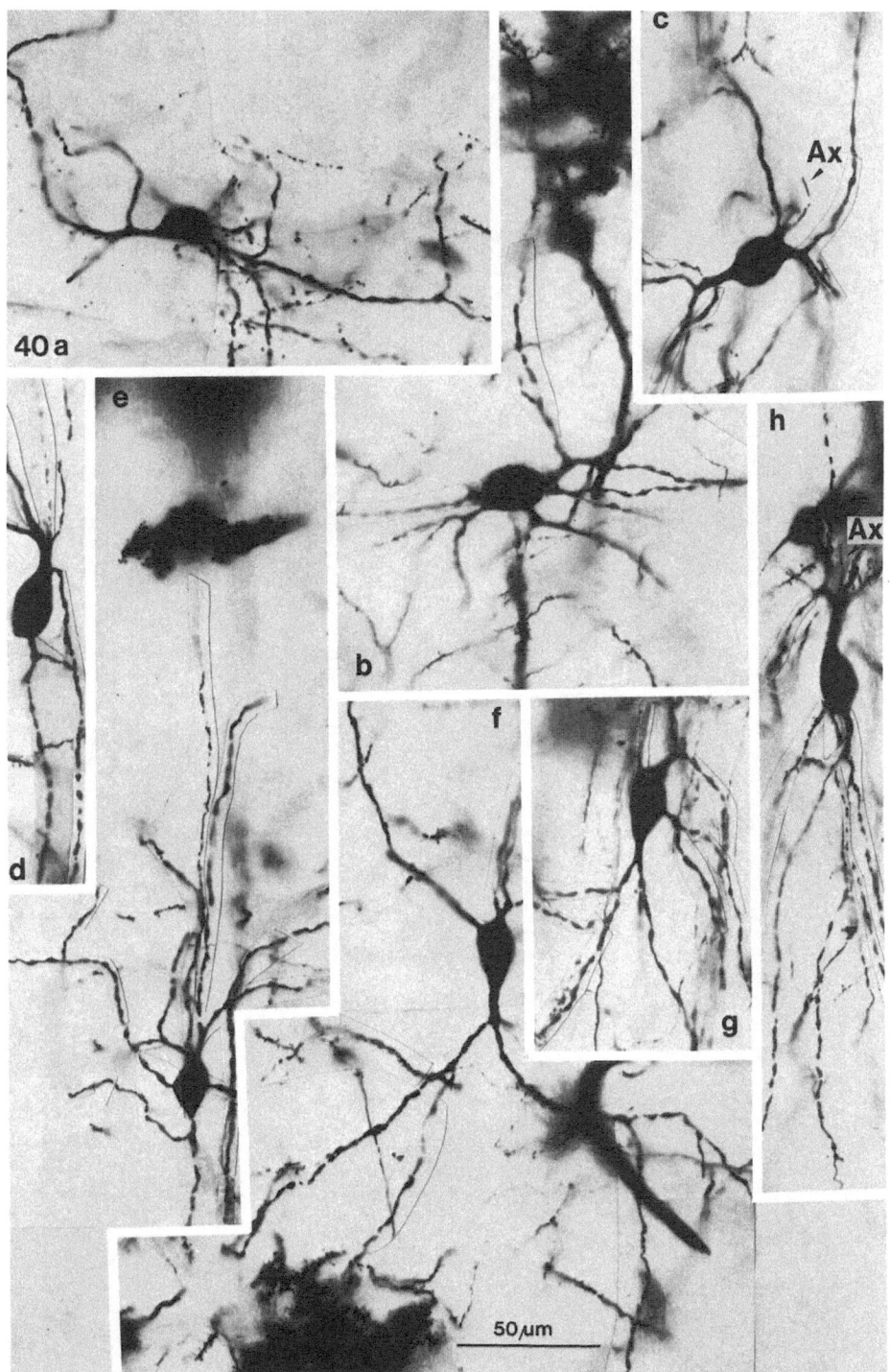

Fig. 40a–h. Photomicrographs of Golgi-impregnated nonpyramidal cells (66-year-old woman). The long axis of the fusiform cell body is oriented horizontally in layer II (*a*), layer Vb (*b*), and layer VIb (*c*), and perpendicular to the cortical surface in layer IIIc/IVa (*e*), layer IVb (*d, g, h*) and layer VIa (*f*). *Ax*, axon

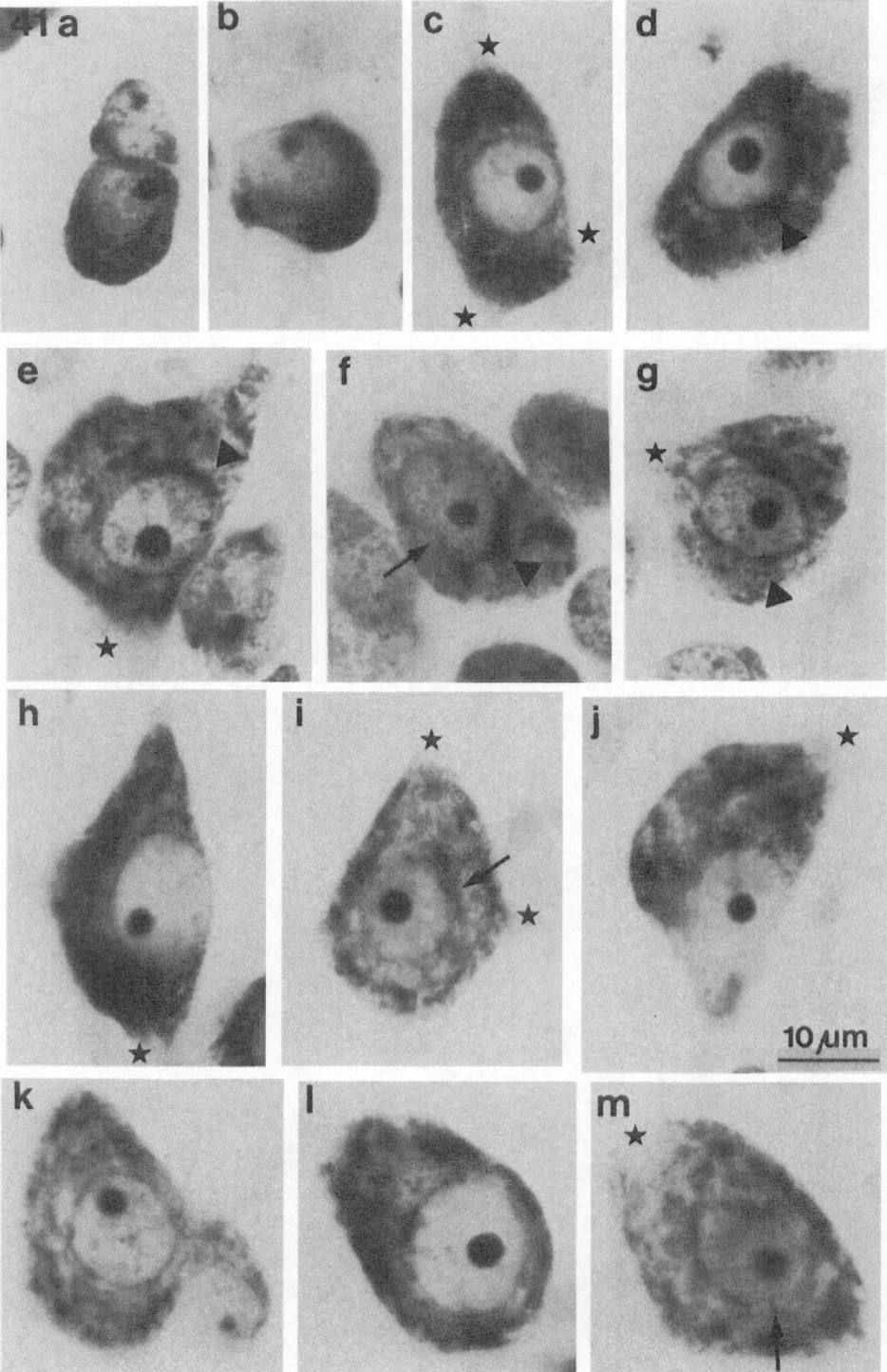

Fig. 41a–m. Nissl-stained large nonpyramidal cells (Araldite, 10 μm, methylene blue; 66-year-old man) of layers II (*a*), IIIab (*b*), IIIc/IVa (*c, d*), IVb (*e*), IVcα (*f, g*), V (*h–k*), and VI (*l, m*). Note the nuclear indentations (*arrows*). Prominent Nissl bodies are marked by *arrowheads*. Frequently the basophilic substance ends, crater-like, at the origin of the dendrites (*asterisks*)

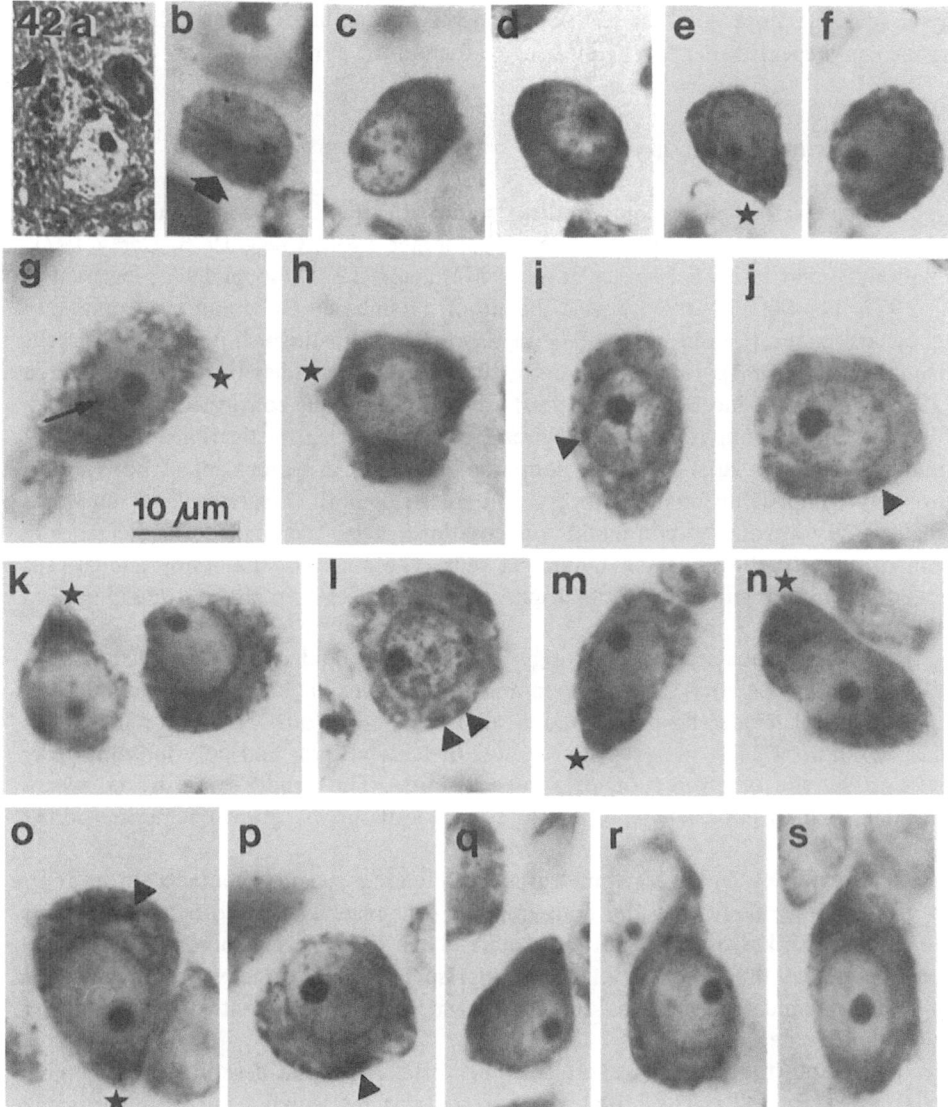

Fig. 42a, b. Photomicrographs of small, pigment-laden nonpyramidal cells of the second layer. One part of the cell body is occupied by the large pigment granules (*short arrow*). *a* Araldite, osmicated, 2 μm, methylene blue-azure II; 66-year-old woman. *b* Araldite, 10 μm, methylene blue; 66-year-old man. The pigment granules are stained in a blue-green shade, hardly distinguishable from the cytoplasm

c–*s* Nissl-stained small nonpyramidal cells (Araldite, 10 μm, methylene blue; 66-year-old man) of layers II (*c*), IIIab (*d*), IIIc/IVa (*e, f*), IVb (*g, i*), IVcα and IVcβ *j, n*), Vb (*o, p*), and VI (*q, s*). Prominent Nissl bodies are marked by *arrowheads*. The basophilic substance usually ends, crater-like, at the origin of the dendrites (asterisks). The *arrow* (*g*) points to a nuclear fold

pyramidal cell, since both the cell body and the nucleus are irregularly shaped. As far as one can judge, it varies to a greater degree than in the pyramidal cell.

11.3 Electron Microscopy

In previous electron-microscopic studies the identification of neurons as nonpyramidal cells was based upon indirect evidence (Colonnier 1968, Cragg 1976, Garey 1971, Lund and Lund 1970, Parnavelas et al. 1977a, Peters 1971, Sloper 1973, Tigges M et al. 1977, Tömböl 1974). Only with the aid of a combined Golgi and electron-microscopic technique (Fairén et al. 1977) is it possible to distinguish nonpyramidal cells unequivocally from pyramidal ones. On the whole, the studies of Peters and Fairén (1978) and Peters and Proskauer (1980) using this technique confirmed the fine structural features of nonpyramidal cells given by the previous investigations. In the adult human striate area large and small nonpyramidal cells can be recognized; both groups probably comprise different cell types. The nuclei of both categories show numerous small, evenly distributed chromatin condensations, giving them a darker appearance in comparison with those of the pyramidal cells. The nuclei of the nonpyramidal cells are commonly irregular in outline, displaying more or less deep indentations (Fig. 43a, b).

In advantageous sections the origin of a dendrite, and in rare cases also its first bifurcation, can be studied. The predominant organelles in the dendrites are microtubules, neurofilaments and mitochondria; ribosomes occur only rarely (see also Peters and Fairén 1978). As an exception clusters of RER cisterns and polyribosomes may exist at the site of bifurcation of a large dendrite. The dendrites are by no means organelle rich as has been reported for stellate cells in the human temporal cortex (Cragg 1976).

The dendrites receive asymmetric and symmetric synaptic contacts. In layer IV of area 17 it is likely that the nonpyramidal cells receive geniculo-cortical afferents along their perikarya and dendrites (Peters et al. 1976a, rat). Somogyi (1978, rat) has shown that the dendrites of the nonpyramidal cells are recipients of the axons of small pyramidal or polygonal nerve cells which are in turn contacted by geniculo-cortical afferents.

In general, the proximal axon does not differ from that described in pyramidal cells (Braak E 1980). The initial segment contains fasciculated microtubules, small rows of smoothly surfaced, spherical and ovoid vesicles (diameter 100–200 nm). In single sections, the initial segment receives no more than three synaptic contacts along a distance of about 10 μm. In each case the axoaxonic synapses are of Gray's type 2.

The soma of the large nonpyramidal cells (Fig. 43b) is rich in organelles when compared with that of the pyramidal cells and small nonpyramidal cells. RER ci-

---

Fig. 43. *a* Electron micrograph of a small nonpyramidal cell of layer IVb (66-year-old woman). Note the eccentric position of the indented nucleus within the cytoplasm; in some areas the organelles accumulate. *b* Electron micrograph of a large nonpyramidal cell of layer Vb (66-year-old woman). The indented nucleus is surrounded by a broad cytoplasmic rim, rich in organelles. Several axosomatic synapses are indicated by *arrows*. *Go*, Golgi complexes; *RER*, RER cisterns

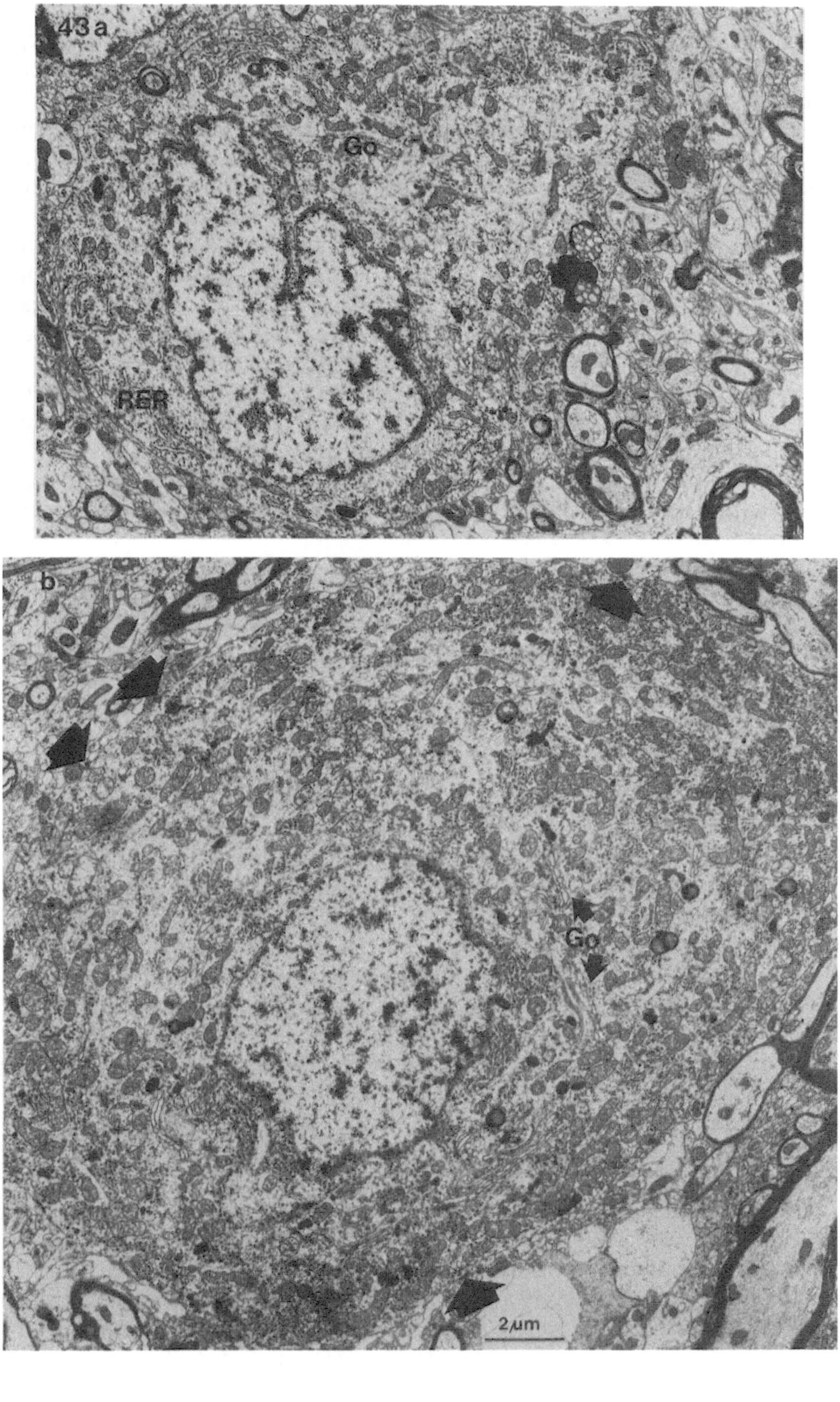

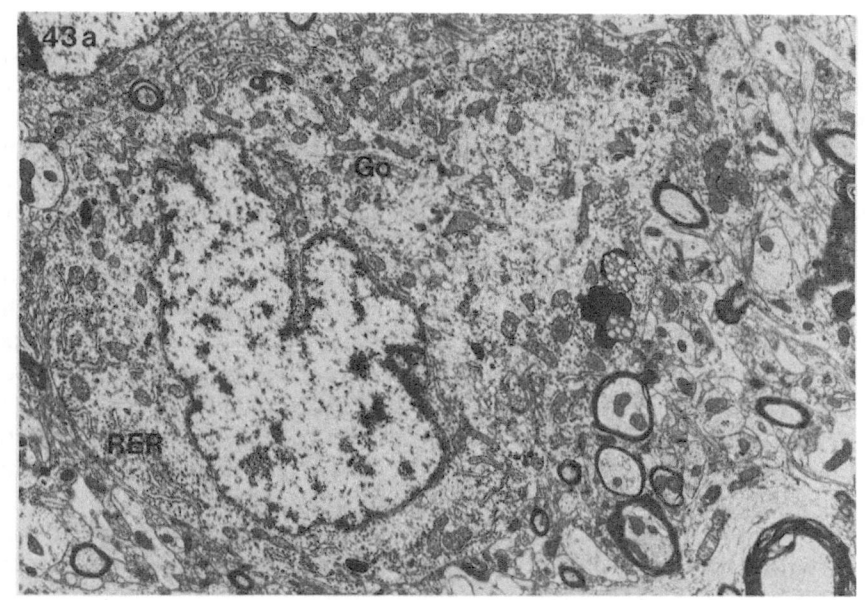

43a

Go

RER

b

Go

2 μm

sterns, ribosomes and mitochondria encompass the nucleus. Beyond that there is often a clearer zone of cytoplasm containing Golgi complexes with remarkably long cisterns. In the peripheral part of the soma RER cisterns and polyribosomes also accumulate. Along the plasmalemma several boutons with spherical vesicles make asymmetric contacts. Close to the specialized postsynaptic membrane and the postsynaptic web, smoothly contoured vesicles, microtubules, dense bodies, and RER cisterns are found. Synapses of the symmetric type rarely occur. In rats they have been shown to originate from other nonpyramidal neurons (Peters and Fairén 1978).

The small nonpyramidal neurons (Fig. 43a) are frequently associated with pyramidal cells. In general, the cytoplasm of these nonpyramidal cells appears somewhat denser than that of the pyramidal cells. Parts of the cytoplasm contain a moderate number of organelles, whereas in others RER cisterns, polyribosomes, and mitochondria accumulate. The RER cisterns are short and branched. Frequently the perinuclear cistern extends into a RER cistern. The mitochondria seem to be somewhat larger than those of the pyramidal cells. Only on very rare occasions do axosomatic synapses occur on the small nonpyramidal cells as compared with the large ones. The ultrastructural features of small, heavily pigmented nonpyramidal cells which occur within layers II and IIIab are in accordance with the same type of neurons in other cortical areas (Braak E 1976, Braak H 1974b).

12 Glial Cells of Layers I to VI

12.1 Astrocytes

Nissl Preparations

Astrocytes are easily identified by means of their lightly stained nuclei with a distinct outline which contrasts with the light cytoplasm (Fig. 3). The nuclei (9–12 × 5–9 μm) of the astrocytes within the upper regions of the cortex show homogeneously distributed chromatin, whereas those in the deeper layers frequently display small heterochromatin clumps. This is probably due to ischemic changes during surgical and fixation procedures (Jenkins et al. 1979). A different vulnerability of astrocytes within different layers should also be taken into account. In the vicinity of the nucleus greenish-yellow-stained pigment inclusions occur.

Astrocytes of the first and upper second layers form the external glial layer (see Braak E 1975). In well preserved tissue the sublayer stains more intensely than the adjoining neuropil (Fig. 2). A faintly tinged external glial layer as well as a "watery" appearance of the vascular glial sheath can be considered a consequence of ischemic changes (Brown and Brierley 1973, Garcia et al. 1977, Kalimo et al. 1977). More than in other layers, the astrocytes of the first layer tend to aggregate in clusters (Janzen 1967, Niessing 1936). The astrocytes of the first layer as well as those of the white matter are termed fibrous astrocytes (for definition see Niessing et al. 1980, pp 30–32); those of the intermediate layers are referred to as protoplasmic astrocytes. In both types, glial fibrillary acidic protein (GFAP) and myosin have been demonstrated by means of an indirect immunofluorescence technique (Braak E et al. 1978, human frontal cortex).

Golgi Preparations

The fibrous astrocytes of the layer I are conspicuous due to the small appendages covering the cell body and the long radiating processes which extend deeply into the layers II and III (Braak E 1975, Braak E et al. 1978, Ramon y Cajal 1909–1911, Retzius 1894a). The protoplasmic astrocytes are often "disturbing" elements in Golgi preparations, since the dense network of their frequently ramifying processes often covers details of impregnated nerve cells (Fig. 6).

Electron Microscopy

Fibrillary and protoplasmic astrocytes (Fig. 44a) can be distinguished from other glial cells by the smooth outline of their nuclei, irregularly shaped cell borders, and a cytoplasm poor in organelles; the mitochondria are characterized by an electron-dense matrix and few cristae (Blinzinger et al. 1965, Braak E 1975, Donelli et al. 1975, Duncan and Morales 1973, Morales and Duncan 1971). There are prominent pigment agglomerations (diameter up to 4 μm) which contain membrane-bound lipid droplets and irregularly shaped electron-dense portions consisting of spherical units of varying size and electron density (Fig. 44b). This astrocytic pigment substructure is different from that of pigment granules found in nerve cells and in oligodendrocytes. It is also different from the pigment found in the astrocytes of 2-year-old rats (Vaughan and Peters 1974). Fibrillary astrocytes exhibit a large number of glial filaments (diameter 8–10 nm) which extend from the perikaryon into the processes. In the protoplasmic astrocytes only a few glial filaments occur. The processes of both types of astrocytes are irregularly contoured and contain all types of cell organelles. The fibrous astrocytes of layer I develop four types of cellular processes, some of which form the external glial layer (Braak E 1975). Gap junctions frequently occur between astroglial processes (Morales and Duncan 1975, Revel and Karnovsky 1967), a fact which may make the exchange of substances between two adjoining astrocytes possible. The surface area of astrocytes constitutes a high percentage of the total surface area of the neuropil (Wolff 1968). Thus, intense interactions with neuronal elements can take place (see Varon and Somjen 1979). The uptake of K^+-ions to maintain the constancy of neuronal environment (Kuffler and Nichols 1966) is at present one of the best-known mechanisms. Furthermore, intense phagocytic activity occurs in consequence of degeneration (Nathaniel and Nathaniel 1977). In end feet of the astrocytes which abut the capillary endothelial cells synthesize considerable quantities of protein and transport it away from the capillary (White 1980, rat).

12.2 Oligodendrocytes

Nissl Preparations

In addition to the typical astrocyte cell nuclei (termed type-I cells in Braak E 1975), intensely stained, irregularly shaped nuclei are present. These nuclei belong to both light oligodendrocytes and microglial cells. Only dark oligodendrocytes can be determined with certainty by the uniformly dense-stained nucleus and cytoplasm. They are the predominating oligodendrocytes. Often one or more oligodendrocytes are associat-

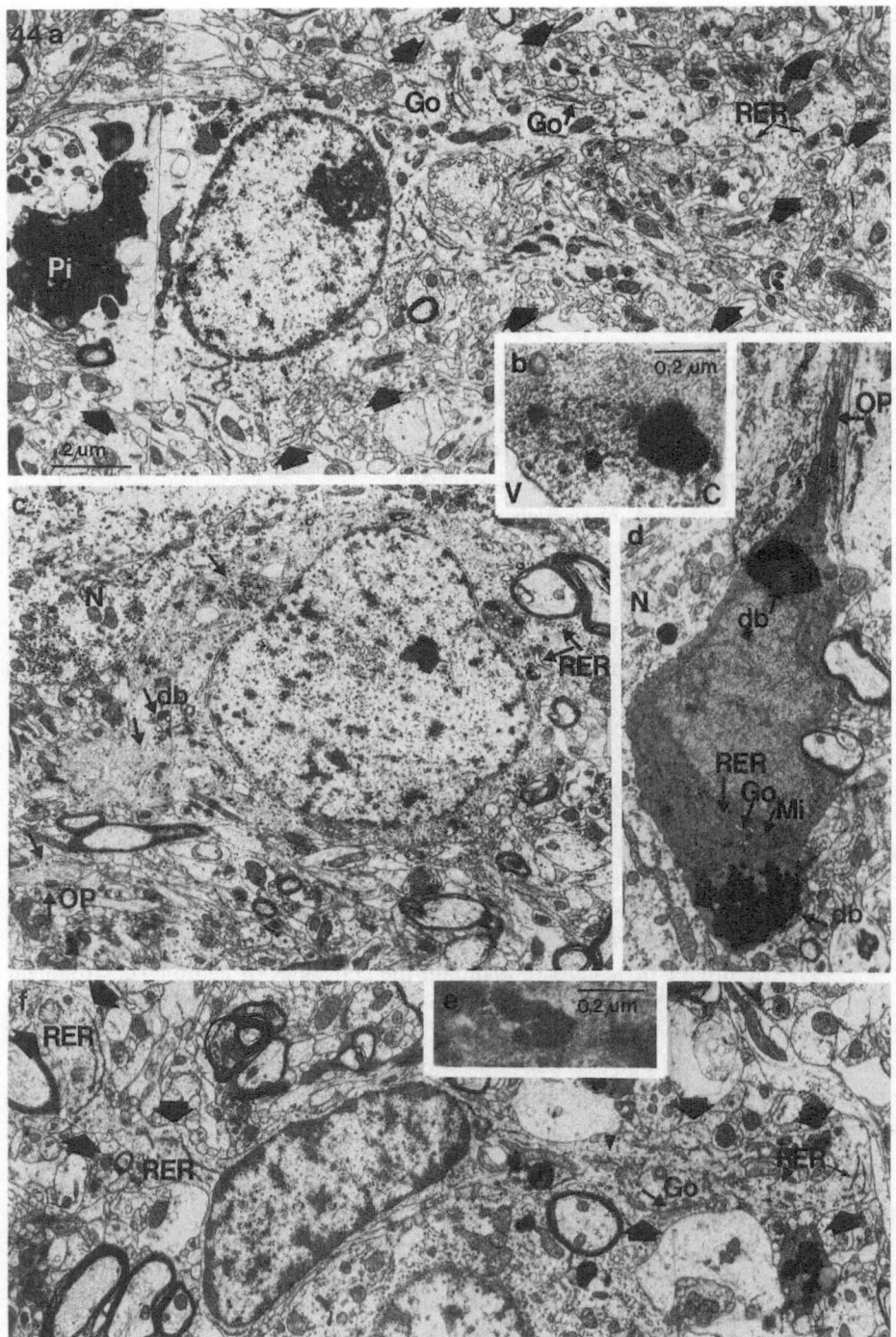

Fig. 44a–f. Electron micrographs of glial cells (66-year-old woman). *Go*, Golgi complexes; *RER*, RER cisterns. *a* Protoplasmic astrocyte with a large agglomeration of pigment (*Pi*). The *arrows* mark the irregularly outlined processes. *b* High magnification of a part of an astrocytic pigment granule consisting of globular subunits of varying electron density. *V*, Lipid vacuole; *C*, cytoplasm. *c* Light oligodendrocyte in close apposition with a nerve cell (*N*). Note the small dense bodies

ed with a nerve cell body. The facing parts of adjacent oligodendrocytes may be flattened.

Electron Microscopy

The dark oligodendrocytes (Fig. 44d) stand out as a result of the remarkable opacity of all parts of the cell. The roughly ovoid or irregularly shaped nucleus is centrally located. Chromatin condensations border the inner nuclear membrane. Mitochondria, cisterns of the endoplasmic reticulum, and microtubules are hardly discernible within the dark, granular cytoplasm. The distinct ribosomes lie in clusters or on the cisterns of the endoplasmic reticulum. The content of the Golgi complexes appears electron lucent. Several microtubules extend into the slender processes. The processes take a straight course and usually lack further organelles. In most of the oligodendrocytes one large pigment inclusion is present (diameter up to 3 μm). Occasionally a membrane-bound lipid droplet is inserted (see also Spoerri et al. 1979, Tusques et al. 1973). Numerous very electron-dense rod-shaped particles (diameter 0.1–0.3 μm) are crowded within an electron-dense matrix. At high magnification these particles reveal a lamellate substructure with a period of 9 nm (Fig. 44e). A pigment of this type cannot be observed in other cells. Neither crystalline inclusions (Vaughan and Peters 1974) nor membranous bodies (Mori and Leblond 1970, Vaughan and Peters 1974, rat) are present in the human oligodendrocytes.

Light oligodendrocytes (Fig. 44c) are rarely found. They can be discriminated from nerve cells, astrocytes, and microglial cells on the basis of the following characteristics: (a) The irregularly shaped nucleus has shallow indentations. Small chromatin condensations adhere to the inner nuclear membrane and are freely dispersed in the nucleoplasm. The perinuclear cistern is uniformly spaced and narrower than in the neighboring nerve cell nuclei. (b) In comparison with nerve cells the contrast of the cytoplasm is slightly enhanced. The mitochondria contain a few round and triangular cristae within a medium electron-dense matrix. Ribosome rosettes and clusters of electron-dense particles – some of them containing an electron-lucent vesicle – are widely scattered. Numerous microtubules are present. (c) Microtubules extend into the cellular processes. The processes take a straight course, tend to have a uniform diameter, and lack RER cisterns and Golgi complexes.

Medium-dense oligodendrocytes, as described by Imamoto et al. (1978), Mori and Leblond (1970), and Mori (1972), have not been observed in the adult human striate area. Mori and Leblond (1970) considered that the light, medium, and dark oligodendrocytes represent a sequential evolution of one type of cell: oligodendroglioblasts give rise to light oligodendrocytes which are transformed into medium ones, and these in turn change into dark oligodendrocytes which may persist for a longer time (Imamoto and Leblond 1978, Ling et al. 1973). The rate of mitosis in glial cells seem

---

(*db*) and the numerous microtubules (*arrows*) which also extend into the processes (*OP*). *d* Dark oligodendrocyte in close apposition with a nerve cell (*N*). The overall electron density of the cytoplasm impedes a clear distinction of the cell organelles, for example the mitochondria (*Mi*). There is pigment accumulation (*db*). The processes (*OP*) take a straight course. The micrograph is tilted 90°. *e* High magnification of an elongated particle of an oligodendrocytic pigment accumulation with its lamellate substructure. *f* Microglial cell with a few microtubules (*arrowheads*) in the irregularly shaped processes marked by *arrows*

71

to be very low during adulthood (Hommes and Leblond 1967; Korr et al. 1973, 1975). In adulthood oligodendrocytes are also capable of phagocytosis (Nathaniel and Nathaniel 1977).

12.3 Microglial Cells

Nissl Preparations

Microglial cells contain a roughly ovoid or crescent-shaped nucleus (about 4–8 μm), accentuated by intensely stained chromatin condensations dispersed within the nucleus and along the nuclear membrane. The narrow cytoplasmic rim is more intensely stained than the surrounding neuropil. One or two short processes emerge from the cell body. Light-microscopically, microglial cells cannot be properly distinguished from light oligodendrocytes.

Electron Microscopy

In silver carbonate impregnations del Rio Hortega (1932) distinguished microglial cells from astrocytes and oligodendrocytes by their cellular processes. Mori and Leblond (1969) and Kitamura et al. (1977) have succeeded in the ultrastructural investigation of silver carbonate impregnated cells; nevertheless, the characteristics of resting microglial cells remain inconsistent. Cells showing the following features are tentatively classified as microglial cells (Fig. 44f): (a) The nucleus is irregularly shaped, lobulated, and smoothly contoured. Heterochromatin forms an almost continuous electron-dense ring along the inner nuclear membrane (Blakemore 1975). (b) The narrow cytoplasmic rim is widened at one cell pole and contains numerous organelles (Vaughan and Peters 1974). (c) One or two irregularly contoured processes extend, which contain all organelles usually found in perikaryal cytoplasm (Blakemore 1975, Mori and Leblond 1969, Vaughan and Peters 1974). (d) Mitochondria, ribosome rosettes, and a few microtubules (Blakemore 1975, Vaughan and Peters 1974), as well as long single strands of RER, are present. These cisterns contain material of a density similar to or higher than the surrounding cytoplasm (Blakemore 1975, Phillips 1973). There are also a few large, electron-dense, membrane-bound particles, perhaps secondary lysosomes, which contain a membranous and tubular (diameter 8 nm) substructure (Fig. 9).

13 Summary

The striate area (Area 17, Brodmann) of the adult human brain (cortical material obtained by surgery) was investigated in Golgi and Nissl preparations and with the electron microscope. Since primary cortical areas receive a defined main input, they are particularly appropriate for clarifying questions concerning intracortical connectivity. The stripe of Gennari is visible to the naked eye. The boundaries of the area can therefore be unequivocally determined. In this study, the nomenclature of the laminae of the striate area proposed by H. Braak (1976) is adopted. Cortical neurons, except those of layer I, can generally be classified as either pyramidal cells or nonpyramidal, i.e., stellate cells.

Layer I. In Golgi preparations, neurons of layer I are very rarely impregnated. There was only one example in which processes could be traced for nearly 100 μm. The processes showed irregularly spaced dilations and constrictions, and terminated in bulbous endings. Clear subdivision of the processes into dendrites and axons was not possible.

In Nissl-stained and methylene blue-azure II-stained sections, layer I stands out with its scarcity of cells. Most of the cells are fibrillary astrocytes, a few belonging to the group of oligodendrocytes and microglial cells. Nerve cells are rarely found. They show a roughly ovoid nucleus surrounded by an intensely stained, narrow cytoplasmic rim. Sometimes a satellite glial cell is visible close to a nerve-cell body.

Electron-microscopically, the nucleus of a layer-I nerve cell generally appears deeply infolded; the cytoplasmic rim is poor in organelles. A few boutons with pleomorphic vesicles can be seen forming axosomatic synaptic contacts. The postsynaptic thickening of these synapses is thin or even absent. In general, the layer-I neurons have features in common with small nonpyramidal cells of the deeper cortical laminae. The neuropil of layer I shows a fairly sharp transition from the membrana limitans gliae superficialis to the subjacent feltwork of processes, the tangential sublamina. This portion of layer I contains numerous tangentially oriented myelinated fibers, intermingled with large synaptic boutons making asymmetric contacts with spines. These boutons resemble those of similar size found in layer IVcβ. On this account they might tentatively be considered geniculate terminals. The tangential sublamina merges gradually into the infratangential sublamina where smaller synaptic boutons can be found. These form asymmetric synaptic contacts with different kinds of dendritic profiles; symmetric synaptic contacts occur only rarely. Almost restricted to the neuropil of layer I are clusters of cross-sectioned small profiles (diameter about 0.1 μm) containing only one or two microtubules. Most of the larger dendritic profiles belong to the terminal tuft of apical dendrites of cortical pyramidal cells.

In Nissl preparations, distinction of cortical laminae is mainly possible because of differences in size and packing density of pyramidal cells. Nonpyramidal, glial, and endothelial cells are more or less randomly dispersed throughout all cortical laminae. The pyramidal cells, in contrast, offer layer-specific characteristics. Before going into detail, it seems appropriate to list features enabling one to classify a cortical nerve cell as either pyramidal or nonpyramidal.

In Golgi preparations, pyramidal cells display a clear polarity with an apical dendrite ascending towards the molecular layer and an axon projecting to the white matter. In general, the dendrites arise gradually from the cell body and taper as they extend distally. They also branch in their distal portions and are more or less densely covered with spines. In nonpyramidal cells the dendrites, as well as the axon, arise at any point from the polymorphic cell body. More often than not, the dendrites originate rather abruptly. The dendrites tend to branch only within the proximal 20–30 μm. Often, at irregular distances, dilations and constrictions can be found. The dendrites lack spines.

In Nissl preparations, pyramidal cells can be identified by the fact that the proximal portions of the apical and basal dendrites contain basophilic substance. The spherical nucleus is centrally located and surrounded by a narrow cytoplasmic rim; the cytoplasm contains evenly distributed basophilic substance. In large neurons Nissl bodies may occur opposite the origins of the apical and basal dendrites. The nonpyramidal cells, in contrast, generally appear more intensely stained. The nucleus is often eccentrically located and infolded, and Nissl bodies are attached to the nuclear indentations. The proximal dendritic segments are devoid of basophilic substance. Often a sharp, crater-like line can be seen at the origin of the dendrites.

Under the electron microscope, pyramidal cells can be identified if an apical dendrite can be traced for about 20–30 μm. The pyramidal cell nucleus is round and smoothly outlined. Within the narrow cytoplasm the Golgi complexes tend to lie in a perinuclear zone. Clusters of cisterns of the rough endoplasmic reticulum (RER) and rosettes of polyribosomes are preferentially located at peripheral parts of the cell body and are also present close to the origins of the dendrites and within the proximal dendritic segments. On rare occasions axosomatic synaptic contacts can be seen; these are of Gray's type 2. In nonpyramidal neurons the infolded nucleus is eccentrically located within a particularly organelle-rich cytoplasm. Clusters of RER cisterns and polyribosomes fill up the nuclear indentations and are widely distributed within the perikaryon. However, they do not penetrate into the proximal dendritic segments. The large nonpyramidal cells bear numerous axosomatic synaptic contacts which are of Gray's type 1 or Gray's type 2. Only the small nonpyramidal cells exhibit a very small number of axosomatic synaptic contacts.

Nonpyramidal Cells of Layers II to VI. Classification of nonpyramidal cells in the adult human striate area is based on shape, size, and orientation of the cell body, as well as on diameter and course of the dendrites. Size and shape of nonpyramidal cell bodies vary considerably. The dendrites arise in three different ways from the cell body: (a) as thick dendritic stems (one or two per cell), which gradually taper as they extend distally and arborize repeatedly at short distances from each other near the soma, (b) as thick dendrites arising abruptly and bifurcating repeatedly near the soma, and (c) as unbranched delicate dendrites. The branching of the dendrites is never profuse. Most of the nonpyramidal cells are either large or small; the diameter of the large ones is about 30 μm. Nonpyramidal cells with an ovoid cell body occur less frequently; their longer axis is either perpendicularly or tangentially oriented. On rare occasions nonpyramidal cells occur with very delicate dendrites or with dendrites curving around the parent soma. In Nissl-stained sections and in ultrathin sections, nonpyramidal cells are either filled with coarse lipofuscin granules or are devoid of pigment. Small, pigment-laden, nonpyramidal cells can be found in great number in layer II and in the upper portions of layer III.

Layer II. Golgi preparations reveal small pyramidal cells with short apical dendrites extending into layer I. Here and there, pyramidal cells show an unusual dendritic arborization with two apical dendrites. Others show the main dendrite originating in lateral parts of the cell body and then taking an ascending course. The size of the cell bodies ranges from 8 to 11 μm. Within the neuropil, radially oriented apical dendrites and dendritic profiles are present. Numerous small boutons with spherical vesicles make asymmetric synaptic contacts with various dendritic profiles.

Layer IIIab. Golgi preparations display medium-sized and small pyramidal cells. In the medium-sized pyramidal cells the apical dendrite gives rise to several apical side branches. It ends in a terminal tuft at the layer II/I border. The small pyramidal cells give off a fine apical dendrite which lacks a terminal tuft. In Nissl preparations, the medium-sized pyramidal cells (diameter 12–15 μm) have a broader cytoplasmic rim than the small ones. A "basophilic cap" is often seen opposite the origin of the apical dendrite. Layer IIIab pyramidal cells are likely to receive input from layer IVcβ neurons. The descending axon of the layer IIIab pyramidal cells branches profusely in layer V before it enters the white matter to form connections with area 18. Within the neurophil, the various types of profiles seen in layer II are still present. In addition, bundles of a few radially oriented myelinated fibers appear.

Layer IV. Layer IV can be divided into several sublayers based on different packing densities of its cellular components. In Golgi preparations, neurons of layer IV offer great variety in shape and size. They cannot easily be classified as pyramidal or nonpyramidal cells; they are termed here "polygonal neurons". A great number of layer-IV cells show spined dendrites, but the dendritic arborization lacks any polarity. On closer examination, these polygonal (multipolar) neurons have more morphological features in common with pyramidal cells than with nonpyramidal cells. They can be considered more or less strongly modified pyramidal cells. The polygonal nerve cells of layer IV show a gradual transition from the soma into the dendrite. The dendrites are covered with spines and also bifurcate in their distal parts. The dendritic diameter gradually reduces from its origin to its distal end. The axon arises from basal parts of the cell body only, and projects towards the white matter. In layer IIIc/IVa polygonal neurons are intermingled with typical pyramidal cells. Here, as well as in the deeper sublayers, are many nerve cells with a pyramid-shaped cell body and with an uncommon appearance of their apical dendrite. This may be very short, tilted, curved, or bifurcated close to the soma, or even lacking. Finally, the polygonal cells do not possess a typical apical dendrite.

In Nissl preparations, the nucleus of the polygonal neurons is spherical and faintly stained. Basophilic substance is predominantly present in the peripheral parts of the soma opposite to the origin of large dendrites and within the proximal dendritic segments.

Under the electron microscope, polygonal neurons — even the large ones — have only a few symmetric axosomatic synaptic contacts. The large polygonal neurons present in layers IIIc/IVa and IVb correspond to the Meynert-Cajal solitary neurons. Layers IIIc/IVa and IVb also contain small and medium-sized polygonal neurons. In experimental animals, the solitary neurons have been found to project to area 18 and other visual areas. Layer IVcα is composed of a mixture of small and medium-sized polygonal neurons. Layer IVcβ, in contrast, displays a homogeneous population of small polygonal neurons with delicate dendrites. In pigment and Nissl preparations, layer IVcβ stands out due to the large number of lipofuscin granules found within the cell bodies of the tiny neurons. The main input to layer IVcβ comes from the parvocellular portions of the lateral geniculate nucleus. In experimental animals, both the geniculate nerve endings and the terminals of the axon collaterals of polygonal layer-IV neurons have been found to contain spherical vesicles. They make asymmetric synaptic contacts.

Layer V. Small and medium-sized polygonal neurons are still present in layer V. Tiny pyramidal cells predominate in the superficial part, layer IVd/Va. They have delicate, unbranched apical dendrites ending in layer IV, sometimes accompanied by ascending axonal collaterals of these cells. The medium-sized pyramidal cells of layer Vb have apical dendrites, also traceable only as far as different levels of layer IV. Side branches are rarely given off from the apical and basal dendrites. The large Meynert pyramidal cells are located mainly in layer Vb. In a few cases their apical dendrite can be seen bifurcating at the layer II/I boundary. Occasionally, similarly large neurons are impregnated, with an apical dendrite bifurcating in layer IV. The apical dendrite gives off side branches only in layer Va. Other dendrites arise from the basal and lateral parts of the cell body; most of them take a horizontal course and can often be traced for about 600 μm. The dendrites bear a few stubby spines, if any at all. The radiate bundles of myelinated fibers piercing layer V are often accompanied by bundles of apical dendrites. The facing dendritic membranes are connected by two types of membrane specialization. Within the neuropil, boutons with pleomorphic vesicles make symmetric synaptic contacts on apical dendrites. Boutons with spherical vesicles have asymmetric contacts on small dendrites and dendritic spines. In experimental animals, lesions in the supragranular layers have been shown to result in degenerating terminals in layer V.

Layer VI. The multiform layer comprises many varieties of nerve cells. Nevertheless, typical features allow one to regard most of them as pyramidal cells. There are small pyramidal cells with an apical dendrite ending in layer Va. They have only a few apical side branches and a few basal dendrites. Medium-sized pyramidal neurons are also present. End ramification of their apical dendrite can be found in layer Va, but one delicate process may ascend into layer IIIc/IVa. Another type of medium-sized pyramidal cell has an apical dendrite which ends without ramification in layer IVb. While transversing layer V it is densely spined and gives off fine, ramifying side branches. Furthermore, there are small and medium-sized triangular neurons characterized by two stout dendrites arising from the cell body opposite to each other or at an obtuse angle. One of these dendrites is radially oriented. Additionally, some thin and rarely branching dendrites may arise from the cell body. All dendritic branches are covered with spines. In addition to these types, there are nerve cells with unusual dendritic arborization, such as endowment with two apical dendrites or a main dendrite arising from lateral parts of the cell body. In experimental animals, layer VI neurons have been demonstrated projecting to layer IV of area 18, to the lateral geniculate body, and to the claustrum.

Nissl preparations display small and medium-sized neurons in layer VI. Especially in the deep tier, neurons occur with a broad cytoplasmic rim. Most of the neurons have a spherical or ovoid nucleus which is smoothly outlined and lightly stained. Occasionally the nucleus offers a nuclear fold.

Astrocytes. Astrocytes can be identified unequivocally. In Nissl preparations they stand out with their lightly and homogeneously stained nuclei. The nuclei are smoothly outlined. Frequently, the cytoplasm of the astrocytes looks lighter than the surrounding neuropil. Lipofuscin pigment agglomerations are present close to the nucleus. In semithin sections the pigment granules are tinged a greenish-yellow shade. In Golgi preparations the fibrillary astrocytes of layer I and of the white matter have long processes following a straight course. The protoplasmic astrocytes, present in layers II to VI, show a great number of short and profusely branching processes. Under the electron microscope the astrocytes display an extremely irregular outline. Both types of astrocytes contain only a few cytoplasmic organelles. Fibrillary and protoplasmic astrocytes can be differentiated by the number of glial filaments: profiles of fibrillary astrocytes contain abundant filaments, whereas profiles of protoplasmic astrocytes often lack filaments. Within the neuropil, astrocytic profiles are identified by mitochondria with an electron-dense granular matrix and few cristae. In the pigment granules, light lipid droplets are embedded in an electron-dense matrix, which consists of densely packed globular units of different size and different electron density.

Oligodendrocytes. Only the dark oligodendrocytes can be determined with certainty in Nissl preparations. Both the cytoplasmic rim and the nucleus are intensely stained. Electron-microscopically, the dark oligodendrocytes can be easily identified by their overall electron opacity. Cell organelles are barely discernible within the cytoplasmic rim: the ribosomes are very dark, and only the luminae of the Golgi complexes appear electron lucent. Large pigment inclusions are frequently present. They consist of an electron-dense matrix with single, small lipid droplets. Numerous electron-dense, rod-shaped particles are crowded within the electron-dense matrix. At high magnification, these particles reveal a lamellate substructure. Oligodendrocytes give rise to a few slender, straight processes. Numerous microtubules extend from the soma into the processes. Occasionally, light oligodendrocytes are found. They display a nucleus with small widespread chromatin condensations and shallow indentations of the nuclear membrane. Small mitochondria with a dense matrix and a few cristae are dispersed within the cytoplasm. Microtubules extend into the cellular processes, which take a straight course. They tend to have a smooth, uniform diameter and lack RER cisterns and Golgi complexes.

Microglial Cells. Glial cells showing the following ultrastructural features are tentatively identified as microglial cells: a lobulated nucleus; heterochromatin clumps forming an almost continuous ring beneath the nuclear membrane; a narrow cytoplasmic rim with a single enlargement encompassing most of the organelles; a cell body with one or more irregularly shaped processes containing all types of organelles usually present within the soma; and RER cisterns, the contents of which appear more electron-dense than the surrounding cytoplasm.

Acknowledgments

The author wishes to express her sincere appreciation to Drs. W. Bargmann †, A.v. Kügelgen † and especially H. Leonhardt for their continual encouragement during her investigations. She gratefully acknowledges Drs. B. Tillmann, K. Unsicker, K. Zilles, R. Lüllmann-Rauch and D. Drenckhahn for their helpful suggestions and criticism. She is particularly grateful to Dr. A.-U. Muhtaroglu (Neurochirurgische Universitätsklinik Kiel, Head: Prof. Dr. H.-P. Jensen) for his kindness in providing surgical material. The author is indebted to Mrs. M. Förster and subsequently to Miss B. Facompré for their skillful technical assistance, to Mrs. S. Piontek for her help with Golgi and pigment preparations, to Mrs. H. Waluk and Mrs. H. Siebke for the photographic work, to Mrs. K. Schmidt-Fichter for correcting the English text, and to Mrs. Graf for typing the manuscript.

Supported by the Deutsche Forschungsgemeinschaft (Br 634/1,2)

References

Allman JM, Kaas JH (1971) A representation of the visual field in the caudal third of the middle temporal gyrus of the owl monkey (Aotus trivirgatus). Brain Res 31:85–105

Allman JM, Kaas JH, Lane RH (1973) The middle temporal visual area (MT) in the bush baby (Galago senegalensis). Brain Res 57:197–202

Anker RL, Cragg BG (1974) Development of the extrinsic connections of the visual cortex in the cat. J Comp Neurol 154:29–42

Armstrong J, Richardson KC, Young JZ (1956) Staining neural end feet and mitochondria after postchromic and carbowax embedding. Stain Technol 31:263–270

Bailey P, Von Bonin G (1951) The isocortex of man. University of Illinois Press, Urbana

Baron M, Gallego A (1971) Cajal cells of the rabbit cortex. Experientia 27:430–432

Beck E (1934) Der Occipitallappen des Affen (Macacus rhesus) und des Menschen in seiner cytoarchitektonischen Struktur. I. Macacus rhesus. J Psychol Neurol (Leipzig) 46:193–323

Benevento LA, Ebner FF (1971) The areas and layers of cortico-cortical terminations in the visual cortex of Virginia opossum. J Comp Neurol 141:157–190

Benevento LA, Rezak M (1976) The cortical projections of the inferior pulvinar and adjacent lateral pulvinar in the rhesus monkey (Macaca mulatta): an autoradiographic study. Brain Res 108:1–24

Benevento LA, Rezak M, Bos J (1975) Extrageniculate projections to layers VI and I of striate cortex (area 17) in the rhesus monkey (Macaca mulatta). Brain Res 96:51–55

Berlin R (1858) Beitrag zur Structurlehre der Großhirnwindungen. Inauguraldissertation, Junge, Erlangen

Bestetti G, Rossi GL (1980) The occurrence of cytoplasmic lamellar bodies in normal and pathological conditions. Acta Neuropathol (Berl) 49:75–78

Billings-Gagliardi S, Chan-Palay V, Palay SL (1974) A review of lamination in area 17 of the visual cortex of Macaca mulatta. J Neurocytol 3:619–629

Blakemore WF (1975) Ultrastructure of normal and reactive microglia. Acta Neuropathol (Berl) [Suppl] VI: 273–278

Blinzinger K, Rewcastle NB, Hager H (1965) Observations on prismatic-type mitochondria within astrocytes of the syrian hamster brain. J Cell Biol 25:293–303

Boothe RG, Greenough WT, Lund JS, Wrege K (1979) A quantitative investigation of spine and dendrite development of neurons in visual cortex (Area 17) of Macaca nemestrina monkeys. J Comp Neurol 186:473–490

Braak E (1975) On the fine structure of the external glial layer of the isocortex of man. Cell Tissue Res 157:367–390

Braak E (1976a) On the fine structure of the small, heavily pigmented non-pyramidal cells in lamina II and upper lamina III of the human isocortex. Cell Tissue Res 169:233–245

Braak E (1976b) Färbung der Nissl-Substanz in 4–10 μm dicken und 2 × 2 cm großen Aralditschnitten. Microsc Acta 78:289–291

Braak E (1978) On the structure of the human striate area. Lamina IVcβ. Cell Tissue Res 188:217–234

Braak E (1980) On the structure of IIIab-pyramidal cells in the human isocortex. A Golgi and electron microscopical study with special emphasis on the proximal axon segment. J Hirnforsch 21:437–442

Braak E, Braak H, Strenge H (1977) The fine structure of myelinated nerve cell bodies in the bulbus olfactorius of man. Cell Tissue Res 182:221–233

Braak E, Drenckhahn D, Unsicker K, Gröschel-Stewart U, Dahl D (1978) Distribution of myosin and the glial fibrillary acidic protein (GFA protein) in rat spinal cord and in the human frontal cortex as revealed by immunofluorescence microscopy. Cell Tissue Res 191:493–499

Braak E, Braak H, Strenge H, Muhtaroglu A-U (1980) Age-related alterations of the proximal axon segment in lamina IIIab-pyramidal cells of the human isocortex. A Golgi and fine structural study. J Hirnforsch 21:531–535

Braak H (1974a) On the structure of the human archicortex. I. The cornu ammonis. A Golgi and pigmentarchitectonic study. Cell Tissue Res 152:349–383

Braak H (1974b) On pigment-loaded stellate cells within layer II and III of the human isocortex. Cell Tissue Res 155:91–104

Braak H (1976) On the striate area of the human isocortex. A Golgi and pigmentarchitectonic study. J Comp Neurol 166:341–364

Braak H (1977) The pigment architecture of the human occipital lobe. Anat Embryol (Berl) 150: 229–250

Braak H (1978) Eine ausführliche Beschreibung pigmentarchitektonischer Arbeitsverfahren. Mikroskopie 34:215–221

Braak H (1980) Architectonics of the human telencephalic cortex. In: Braitenberg V, Barlow HB, Florey E, Grüsser OJ, Van der Loos H (eds) Studies of brain function, vol 4. Springer, Berlin Heidelberg New York

Braak H, Braak E, Strenge H (1976) Gehören die Inselneurone der Regio entorhinalis zur Klasse der Pyramiden oder der Sternzellen? Z Mikrosk Anat Forsch 90:1017–1031

Bradford R, Parnavelas JG, Lieberman AR (1977) Neurons in layer I of the developing occipital cortex of the rat. J Comp Neurol 176:121–132

Braitenberg V, Guglielmotti V, Sada E (1967) Correlation of crystal growth with the staining of axons by the Golgi procedure. Stain Technol 42:277–283

Brodmann K (1903) Beiträge zur histologischen Lokalisation der Großhirnrinde. II. Mitteilung: Der Calcarinatypus. J Psychol Neurol (Leipzig) 2:133–159

Brodmann K (1909) Vergleichende Lokalisationslehre der Großhirnrinde. Barth JA, Leipzig

Brown AW, Brierley JB (1973) The earliest alterations in rat neurons and astrocytes after anoxia-ischaemia. Acta Neuropathol (Berl) 23:9–22

Buschmann MBT (1979) Development of lamellar bodies and subsurface cisterns in pyramidal cells and neuroblasts of hamster cerebral cortex. Am J Anat 155:175–184

Butler AB, Jane JA (1977) Interlaminar connections of rat visual cortex: an ultrastructural study. J Comp Neurol 174.521–534

Butler AB, Jane JA, Falk P (1979) Interlaminar connections of the tree shrew visual cortex. Neurosci Lett 11.107–110

Carey RG, Fitzpatrick D, Diamond IT (1979) Layer I of striate cortex of Tupaia glis and Galago senegalensis: projections from thalamus and claustrum revealed by retrograde transport of horseradish peroxidase. J Comp Neurol 186:393–438

Carey RG, Bear MF, Diamond IT (1980) The laminar organization of the reciprocal projections between the claustrum and striate cortex in the tree shrew, Tupaia glis. Brain Res 184:193–198

Chan-Palay V, Palay SL, Billings-Gagliardi SM (1974) Meynert cells in the primate visual cortex. J Neurocytol 3:631–658

Clark WE, LeGros (1942) The cells of Meynert in the visual cortex of the monkey. J Anat 76: 369–476

Clark WE, LeGros, Sunderland S (1939) Structural changes in the isolated visual cortex. J Anat 73:563–574

Colonnier M (1964) Experimental degeneration in the cerebral cortex. J Anat 98:47–54

Colonnier M (1966) The structural design of the neocortex. In: Eccles JC (ed) Brain and conscious experience. Springer, Berlin Heidelberg New York, pp 1–23

Colonnier M (1967) The fine structural arrangement of the cortex. Arch Neurol 16:651–657

Colonnier M (1968) Synaptic patterns on different cell types in the different laminae of the cat visual cortex. An electron microscope study. Brain Res 9:268–287

Colonnier M, Sas E (1978) An anterograde degeneration study of the tangential spread of axons in cortical areas 17 and 18 of the squirrel monkey (Saimiri sciureus). J Comp Neurol 179:245–262

Conel JL (1939, 1941, 1947, 1951, 1955, 1959, 1963, 1967) The postnatal development of the human cerebral cortex, vol I-VIII. Harvard University Press, Cambridge MA

Cragg BG (1976) Ultrastructural features of human cerebral cortex. J Anat 121:331–362

del Rio Hortega P (1932) Microglia. In: Penfield W (ed) Cytology and cellular pathology of the nervous system, vol. 2. PB Hoeber, New York, pp 483–534

Donelli G, D'Uva V, Paoletti L (1975) Ultrastructure of gliosomes in ependymal cells of the lizard. J Ultrastruct Res 50:253–263

Doty RN, Kimura DS, Mogenson GJ (1964) Photically and electrically elicited responses in the central visual system of the squirrel monkey, Saimiri sciureus. Exp Neurol 10:19–51

Duncan D, Morales R (1973) Fine structure of astrocyte mitochondria in the spinal cord of the dog, cat and monkey. Anat Rec 175:519–528

Fairén A, Peters A, Saldanha J (1977) A new procedure for examining Golgi impregnated neurons by light and electron microscopy. J Neurocytol 6:311–337

Feldman ML, Peters A (1978) The forms of non-pyramidal neurons in the visual cortex of the rat. J Comp Neurol 179:761–794

Ferster D, LeVay S (1978) The axonal arborization of lateral geniculate neurons in the striate cortex of the cat. J Comp Neurol 182:923–944

Filimonoff IN (1932) Über die Variabilität der Großhirnrindenstruktur. II. Regio occipitalis beim erwachsenen Menschen. J Psychol Neurol (Leipzig) 44:1–96

Fisken RA, Garey LJ, Powell TPS (1975) The intrinsic, association and commissural connections of area 17 of the visual cortex. Philos Trans R Soc Lond Biol 272:487–535

Fleischhauer K (1974) On different patterns of dendritic bundling in the cerebral cortex of the cat. Z Anat Entwicklungsgesch 143:115–126

Fleischhauer K, Laube A (1977) A pattern formed by preferential orientation of tangential fibres in layer I of the rabbit cerebral cortex. Anat Embryol (Berl) 151:233–240

Fleischhauer K, Vossel A (1979) Cell densities in the various layers of the rabbit's striate area. Anat Embryol (Berl) 156:269–281

Fleischhauer K, Petsche H, Wittkowski W (1972) Vertical bundles of dendrites in the neocortex. Z Anat Entwicklungsgesch 136:213–223

Fox MW, Inman O (1966) Persistence of Retzius-Cajal cells in the developing dog brain. Brain Res 3:192–194

Fox MW, Inman OR, Himwich WA (1966) The postnatal development of neocortical neurons in the dog. J Comp Neurol 127:199–206

Garcia JH, Kalimo H, Kamijyo J, Trump BF (1977) Cellular events during partial cerebral ischemia. I. Electron microscopy of feline cerebral cortex after middle-cerebral-artery occlusion. Virchows Archiv [Cell Pathol] 25:191–206

Garcia JH, Lossinsky AS, Kauffman FC, Conger KA (1978) Neuronal ischemic injury: light microscopy, ultrastructure and biochemistry. Acta Neuropathol (Berl) 43:85–95

Garey LJ (1971) A light and electron microscopic study of the visual cortex of the cat and monkey. Proc R Soc Lond [Biol] 179:21–40

Garey LJ, Powell TPS (1971) An experimental study of the termination of the lateral geniculo-cortical pathway in the cat and monkey. Proc R Soc Lond [Biol] 179:41–63

Gilbert CD, Kelly JP (1975) The projection of cells in different layers of the cat visual cortex. J Comp Neurol 163:81–106

Gilbert CD, Wiesel TN (1979) Morphology and intracortical projections of functionally characterized neurones in the cat visual cortex. Nature 280:120–126

Glendenning KK, Kofron EA, Diamond IT (1976) Laminar organization of the lateral geniculate nucleus to the striate cortex in Galago. Brain Res 105:538–546

Globus A, Scheibel AB (1967) Pattern and field in cortical structure: the rabbit. J Comp Neurol 131:155–172

Gray EG (1959) Axosomatic and axodendritic synapses of the cerebral cortex: an electron microscopic study. J Anat 93:420–433

Hassler R, Wagner A (1965) Experimentelle und morphologische Befunde über die vierfache Projektion des visuellen Systems. In: Proceedings of the 8th international congress for neurology, 5–10 Sept 1965. Disturbances of the Occipital Lobe. III, 77–96, Wien 1965. Ref.: Excerpta Med. Intern. Congr. Ser. Nr. 94, 24–26

Hendrickson AE, Wilson JR, Ogren MP (1978) The neuroanatomical organization of pathways between the dorsal lateral geniculate nucleus and visual cortex in old-world and new-world primates. J comp Neurol 182:123–136

Herndon RM (1964) The fine structure of the rat cerebellum: II. The stellate neurons, granule cells and glia. J Cell Biol 23:277–293

Holländer H (1974) Projections from the striate cortex to the diencephalon in the squirrel monkey (Saimiri sciureus). A light microscopic radioautographic study following intracortical injection of ^{3}H-leucine. J Comp Neurol 155:425–440

Hommes OR, Leblond CP (1967) Mitotic division of neuroglia in the normal rat. J Comp Neurol 129:269–278

Hubel DH, Wiesel TN (1962) Receptive fields, binocular interaction and functional architecture in cat's visual cortex. J Physiol 160:106–154

Hubel DH, Wiesel TN (1972) Laminar and columnar distribution of geniculo-cortical fibers in the macaque monkey. J Comp Neurol 146:421–450

Hubel DH, Wiesel TN (1977) Ferrier lecture. Functional architecture of macaque monkey visual cortex. Proc R Soc Lond [Biol] 198:1–59

Imamoto K, Leblond CP (1978) Radioautographic investigation of gliogenesis in the corpus callosum of young rats. II. Origin of microglial cells. J Comp Neurol 180:139–164

Imamoto K, Paterson JA, Leblond CP (1978) Radioautographic investigation of gliogenesis in the corpus callosum of young rats. I. Sequential changes in oligodendrocytes. J Comp Neurol 180:115–138

Janzen RWC (1967) Topographische Besonderheiten im Bau der Glia marginalis des Menschen. Z Zellforsch 80:570–584

Jenkins LW, Povlishock JT, Becher DP, Miller JD, Sullivan HG (1979) Complete cerebral ischemia. An ultrastructural study. Acta Neuropathol (Berl) 48:113–125

Jones EG (1975) Varieties and distribution of non-pyramidal cells in the somatic sensory cortex of the squirrel monkey. J Comp Neurol 160:205–267

Jones EG, Powell TPS (1970) Electron microscopy of the somatic sensory cortex in the cat. I. II. III. Philos Trans R Soc Lond [Biol] 257:1–28

Kalimo H, Garcia JH, Kamijyo Y, Tanaka J, Trump BF (1977) The ultrastructure of brain death. II. Electron microscopy of feline cortex after complete ischemia. Virchows Archiv [Cell. Pathol] 25:207–220

Karnovsky MJ (1967) The ultrastructural basis of capillary permeability studied with peroxidase as a tracer. J Cell Biol 35:213–236

Kelly JP, Van Essen DC (1974) Cell structure and function in the visual cortex of the cat. J Physiol (Lond) 238:515–547

Kirsche W, Kunz G, Wenzel J, Wenzel M, Winkelmann E, Winkelmann A (1973) Neurohistologische Untersuchungen zur Variabilität der Pyramidenzellen des sensorischen Cortex der Ratte. J Hirnforsch 14:117–135

Kitamura T, Tsuchihashi Y, Tatebe A, Fujita S (1977) Electron microscopic features of the resting microglia in the rabbit hippocampus, identified by silver carbonate staining. Acta Neuropathol (Berl) 38:195–201

König N, Valat J, Fulcrand J, Marty R (1977) The time of origin of Cajal-Retzius cells in the rat temporal cortex. An autoradiographic study. Neurosci Lett 4:21–26

Korr H, Schultze B, Maurer W (1973) Autoradiographic investigations of glial proliferation in the brain of adult mice. I. The DNA synthesis phase of neuroglia and endothelial cells. J Comp Neurol 150:169–176

Korr H, Schultze B, Maurer W (1975) Autoradiographic investigations of glial proliferation in the brain of adult mice. II. Cycle time and mode of proliferation of neuroglia and endothelial cells. J Comp Neurol 160:477–490

Kuffler SW, Nicholls JG (1966) The physiology of neuroglial cells. Ergeb Physiol 57:1–90

LeBeux YJ (1972) Subsurface cisterns and lamellar bodies: particular forms of the endoplasmic reticulum in the neuron. Z Zellforsch 133:327–352

LeVay S (1973) Synaptic patterns in the visual cortex of the cat and monkey. Electron microscopy of Golgi preparations. J Comp Neurol 150:53–86

Levey NH, Jane JA (1975) Laminar thermocoagulation of the visual cortex of the rat. Brain Behav Evol 11:257–274

Lin CS, Kaas JH (1977) Projections from cortical visual areas 17, 18, and MT onto the dorsal lateral geniculate nucleus in owl monkey. J Comp Neurol 173:457–474

Lin CS, Friedländer MJ, Sherman SM (1979) Morphology of physiologically identified neurons in the visual cortex of the cat. Brain Res 172:344–348

Ling EA, Paterson JA, Privat A, Mori S, Leblond CP (1973) Investigation of glial cells in semithin sections. Identification of glial cells in the brain of young rats. J Comp Neurol 149:43–72

Lopes CAS, Mair WGP (1974) Ultrastructure of the outer cortex and the pia mater in man. Acta Neuropathol (Berl) 28:79–86

Lorente de Nó R (1934) Studies on the structure of the cerebral cortex. II. Continuation of the study of the ammonic system. J Psychol Neurol (Leipzig) 46:113–177

Lorente de Nó R (1938) The cerebral cortex: architecture, intracortical connections and motor projections. In: Fulton JF (ed) Physiology of the nervous system. Oxford University Press, London New York Toronto, pp 291–321

Lund JS (1973) Organization of neurons in the visual cortex, area 17, of the monkey (Macaca mulatta). J Comp Neurol 147:455–496

Lund JS, Boothe RG (1975) Interlaminar connections and pyramidal neuron organisation in the visual cortex, area 17, of the macaque monkey. J Comp Neurol 159:305–334

Lund JS, Lund RD (1970) The termination of callosal fibres in the paravisual cortex of the rat. Brain Res 17:25–45

Lund JS, Lund RD, Hendrickson AE, Bunt AH, Fuchs AF (1975) The origin of efferent pathways from the primary visual cortex, area 17, of macaque monkey as shown by retrograde transport of horseradish peroxidase. J Comp Neurol 164:287–304

Lund JS, Boothe RG, Lund RD (1977) Development of neurons in the visual cortex (area 17) of the monkey (Macaca nemestrina): A Golgi study from fetal day 127 to postnatal maturity. J Comp Neurol 176:149–188

Magalhães-Castro HH, Saraiva PES, Magalhães-Castro B (1975) Identification of corticotectal cells of the visual cortex of cats by means of horseradish peroxidase. Brain Res 83:474–479

Marin-Padilla M (1972) Prenatal ontogenetic history of the principal neurons of the neocortex of the cat (Felis domestica). A Golgi study. II. Developmental differences and their significances. Z Anat Entwicklungsgesch 136:125–142

Martinez-Millán L, Holländer H (1975) Cortico-cortical projections from striata cortex of the squirrel monkey (Saimiri sciureus). A radioautographic study. Brain Res 83:405–417

Maurer J, Fleischhauer K (1979) Preferential orientation of small profiles in neuropil of lamina I. A quantitative ultrastructural study of tangential sections through sublamina tangentialis of rabbit visual cortex. Anat Embryol (Berl) 157:133–149

Meller K, Breipohl W, Glees P (1969) Ontogeny of the mouse motor cortex. The polymorph layer or layer VI. A Golgi and electronmicroscopical study. Z Zellforsch 99:443–458

Meynert T (1867) Der Bau der Großhirnrinde und seine örtlichen Verschiedenheiten, nebst einem pathologisch-anatomischen Corollarium. V Jahresschr Psychiatr 1:77–93

Meynert T (1868) Der Bau der Großhirnrinde und seine örtlichen Verschiedenheiten, nebst einem pathologisch-anatomischen Corollarium. V Jahresschr Psychiatr 2:88–113

Morales R, Duncan D (1971) Prismatic and other unusual arrays of mitochondria cristae in astrocytes of cats and hamsters. Anat Rec 171:545–558

Morales R, Duncan D (1975) Specialized contacts of astrocytes with astrocytes and with other cell types in the spinal cord of the cat. Anat Rec 182:255–266

Mori S (1972) Ligth and electron microscopic features and frequencies of the glial cells present in the cerebral cortex of the rat brain. Arch Histol Jpn 34:231–244

Mori S, Leblond CP (1969) Identification of microglia in light and electron microscopy. J Comp Neurol 135:57–80

Mori S, Leblond CP (1970) Electron microscopic identification of three classes of oligodendrocytes and a preliminary study of their proliferative activity in the corpus callosum of young rats. J Comp Neurol 139.1–30

Nathaniel EJJ, Nathaniel DR (1977) Astroglial response to degeneration of dorsal root fibres in adult spinal cord. Exp Neurol 54:60–76

Nauta HJW, Butler AB, Jane JA (1973) Some observations on axonal degeneration resulting from superficial lesions of the cerebral cortex. J Comp Neurol 150:349–360

Niessing K (1936) Über systemartige Zusammenhänge der Neuroglia im Großhirn und über ihre funktionelle Bedeutung. Gegenbaurs Morphol Jahrb 78:537–584

Niessing K, Scharrer E, Scharrer B, Oksche A (1980) Die Neuroglia: Historischer Überblick. In: Oksche A, Vollrath L (eds) Handbuch der mikroskopischen Anatomie des Menschen, vol 4/10. Springer, Berlin Heidelberg New York, pp 1–156

Noback CR, Purpura DP (1961) Postnatal ontogenesis of neurons in cat neocortex. J Comp Neurol 117:291–308

Ogren M, Hendrickson A (1976) Pathway between striate cortex and subcortical regions in Macaca mulatta and Saimiri sciureus: evidence for a reciprocal pulvinar connection. Exp Neurol 53:780–800

82

Ogren MP, Hendrickson AE (1977) The distribution of pulvinar terminals in visual areas 17 and 18 of the monkey. Brain Res 137:343–350

Ogren MP, Hendrickson AE (1979) The morphology and distribution of striate cortex terminals in the inferior and lateral subdivision of the Macaca monkey pulvinar. J Comp Neurol 188:200

Palay SL, Sotelo C, Peters A, Orkand PM (1968) The axon hillock and the initial segment. J Cell Biol 38.193–201

Parnavelas JG, Sullivan K, Lieberman AR, Webster KE (1977a) Neurons and their synaptic organization in the visual cortex of the rat. Electron microscopy of Golgi preparations. Cell Tissue Res 183:499–517

Parnavelas JG, Lieberman AR, Webster KE (1977b) Organization of neurons in the visual cortex, area 17, of the rat. J Anat 124:305–322

Peters A (1971) Stellate cells of the rat parietal cortex. J Comp Neurol 141:345–374

Peters A, Fairén A (1978) Smooth and sparsely-spined stellate cells in the visual cortex of the rat: a study using a combined Golgi-electron microscope technique. J Comp Neurol 181:129–172

Peters A, Feldman ML (1976) The projection of the lateral geniculate nucleus to area 17 of the rat cerebral cortex. I. General description. J Neurocytol 5:63–84

Peters A, Feldman ML (1977) The projection of the lateral geniculate nucleus to area 17 of the rat cerebral cortex. IV. Terminations upon spiny dendrites. J Neurocytol 6:669–689

Peters A, Kaiserman-Abramof IR (1970) The small pyramidal neuron of the rat cerebral cortex. The perikaryon, dendrites and spines. Am J Anat 127:321–356

Peters A, Proskauer CC (1980) Smooth and sparsely spined cells with myelinated axons in rat visual cortex. Neuroscience 5:2079–2092

Peters A, Saldanha J (1976) The projection of the lateral geniculate nucleus to area 17 of the rat cerebral cortex. III. Layer VI. Brain Res 105:533–537

Peters A, Walsh TM (1972) A study of the organization of apical dendrites in the somatic sensory cortex of the rat. J Comp Neurol 144:253–268

Peters A, Proskauer CC, Kaiserman-Abramof IR (1968) The small pyramidal neuron of the rat cerebral cortex. The axon hillock and initial segment. J Cell Biol 39:604–619

Peters A, Feldman M, Saldanha J (1976a) The projection of the lateral geniculate nucleus to area 17 of the rat cerebral cortex. II. Terminations upon neuronal perikarya and dendritic shafts. J Neurocytol 5:85–107

Peters A, Palay SL, Webster HdeF (1976b) The fine structure of the nervous system: the neurons and supporting cells. WB Saunders, Philadelphia

Peters A, Proskauer CC, Feldman ML, Kimerer L (1979) The projection of the lateral geniculate nucleus to area 17 of the rat cerebral cortex. V. Degenerating axon terminals synapsing with Golgi impregnated neurons. J Neurocytol 8:331–357

Phillips DE (1973) An electron microscopic study of macroglia and microglia in the lateral funiculus of the developing spinal cord in the fetal monkey. Z Zellforsch 140:145–167

Raczkowski D, Diamond IT (1978) Connections of the striate cortex in Galago senegalensis. Brain Res 144:383–388

Ramón y Cajal S (1890) Textura de las circonvoluciones cerebrales de los mamiferos inferiores. Nota preventiva. Gac Medica Catal December 15, 1890

Ramón y Cajal S (1891) Sur la structure de l'écorce cérébrale de quelques mammiféres. Cellule 7:125–176

Ramón y Cajal S (1893) Nuevo concepto de la histologia de los centros nerviosos. Rev Cienc Med Barcelona

Ramón y Cajal S (1900) Studien über die Hirnrinde des Menschen. 1. Die Sehrinde. Barth, Leipzig

Ramón y Cajal S (1909–1911) Histologie du système nerveux de l'homme et des vertébrés. Maloine, Paris (Reprinted 1952–1955 by Consejo superior de Investigaciones cientificas, Madrid)

Ramón y Cajal S (1929) Studies on vertebrate neurogenesis. Translated by Guth L (1959). CC Thomas, Springfield IL

Ramsey HJ (1965) Fine structure of the surface of the cerebral cortex of human brain. J Cell Biol 26:323–333

Retzius G (1891) Über den Bau der Oberflächenschicht der Großhirnrinde des Menschen und bei Säugetieren. Verh Biol Ver Stockh 1

Retzius G (1893) Die Cajal'schen Zellen der Großhirnrinde beim Menschen und bei Säugetieren. Biol Untersuch [Neue Folge] 5:1–8

Retzius G (1894a) Die Neuroglia des Gehirns beim Menschen und bei Säugetieren. Biol Untersuch [Neue Folge] 6:1–28

Retzius G (1894b) Weitere Beiträge zur Kenntnis der Cajal'schen Zellen der Großhirnrinde des Menschen. Biol Untersuch [Neue Folge] 6:29–34

Revel JP, Karnovsky MJ (1967) Hexagonal array of subunits in intercellular junctions of the mouse heart and liver. J Cell Biol 33:C7–C12

Reyners H, Gianfelici de Reyners E, Van der Parren J, Maisin J-R (1977) Etude morphométrique des cisternes submembranaires des neurones et de leurs relations spatiales avec les processus astrocytaires dans le cortex cérébral du rat. Biol Cell 30:265–278

Reynolds ES (1963) The use of lead citrate at high pH as an electronopaque stain in electron microscopy. J Cell Biol 17:208–211

Rezak M, Benevento LA (1979) A comparison of the organization of the projections of the dorsal lateral geniculate nucleus, the inferior pulvinar and adjacent lateral pulvinar to primary visual cortex (area 17) in the macaque monkey. Brain Res 167:19–40

Rhoades RW, Chalupa LM (1978) Functional and anatomical consequences of neonatal visual cortical damage in superior colliculus of the golden hamster. J Neurophysiol 41:1466–1494

Ribak CE, Peters A (1975) An autoradiographic study of the projections from the lateral geniculate body of the rat. Brain Res 92:341–368

Richardson KC, Jarett L, Finke EH (1960) Embedding in epoxy resins for ultrathin sectioning in electron microscopy. Stain Technol 35:313–323

Rickmann M, Chronwall BM, Wolff JR (1977) On the development of non-pyramidal neurons and axons outside the cortical plate: the early marginal zone as a pallial anlage. Anat Embryol 151:285–307

Robson JA, Hall WC (1975) Connections of layer VI in striate cortex of the grey squirrel (Sciureus carolinensis). Brain Res 93:133–139

Rockland KS, Pandya DN (1979) Laminar origins and terminations of cortical connections of the occipital lobe in the rhesus monkey. Brain Res 179:3–20

Rosenbluth J (1962) Subsurface cisterns and their relationship to the neuronal plasma membrane. J Cell Biol 13:405–421

Rosenquist AC, Edwards SB, Palmer LA (1974) An autoradiographic study of the projections of the dorsal lateral geniculate nucleus and the posterior nucleus in the cat. Brain Res 80:71–93

Rowe MH, Benevento LA, Rezak M (1978) Some observations on the patterns of segregated geniculate inputs to the visual cortex in new world primates: an autoradiographic study. Brain Res 159:371–378

Sanides F, Vitzthum H (1965) Die Grenzerscheinungen am Rande der menschlichen Sehrinde. Dtsch Z Nervenheilkd 187:708–719

Sas E, Sanides F (1970) A comparative Golgi study of Cajal foetal cells. Z Mikrosk Anat Forsch 82:385–396

Schierhorn H, Doedens K, Nagel I (1973) Über das spine-freie („nackte") Initialsegment der Apikaldendriten von corticalen Pyramidenzellen der Albinoratte. Gegenbaurs Morphol Jahrb 119: 130–145

Schober W, Winkelmann E (1977) Die geniculo-kortikale Projektion bei Albinoratten. J Hirnforsch 18:1–20

Schober W, Lüth HJ, Gruschka H (1976) Die Herkunft afferenter Axone im striären Kortex der Albinoratte: eine Studie mit Meerrettichperoxidase. Z Mikrosk Anat Forsch 90:399–415

Shkol'nik-Yarros EG (1971) Neurons and interneuronal connections of the central visual system. Plenum Press, New York

Sholl DA (1956) The organization of the cerebral cortex. Methuen, London

Sloper JJ (1973) An electron microscopic study of the neurons of the primate motor and somatic sensory cortices. J Neurocytol 2:351–358

Smith DE, Moskowitz N (1978) Ultrastructure of layer IV of the primary auditory cortex of the squirrel monkey. Neuroscience 4:349–359

Somogyi P (1978) The study of Golgi stained cells and of experimental degeneration under the electron microscope: a direct method for the identification in the visual cortex of three successive links in a neuron chain. Neuroscience 3:167–180

Sotelo C (1969) Ultrastructural aspects of the cerebellar cortex of the frog. In: Llinas R (ed) Neurobiology of cerebellar evolution and development. AMA Education and Research Foundation, Chicago, pp 327–371

Sousa-Pinto A, Paula-Barbosa M, DoCarmoMatos M (1975) A Golgi and electron microscopical study of nerve cells in layer I of the cat auditory cortex. Brain Res 95:443–458

Spatz WB (1975) An efferent connection of the solitary cells of Meynert. A study with horseradish peroxidase in the marmoset Callithrix. Brain Res 92:450–455

Spatz WB (1977) Topographically organized reciprocal connections between area 17 and MT (visual area of superior temporal nucleus) in the marmoset Callithrix jacchus. Exp Brain Res 27:559–572

Spatz WB (1979) The retino-geniculo-cortical pathway in Callithrix. II. The geniculo-cortical projection. Exp Brain Res 36:401–410

Spatz WB, Tigges J (1972) Experimental-anatomical studies on the middle temporal visual area (MT) in primates. I. Efferent cortico-cortical connection in the marmoset Callithrix jacchus. J Comp Neurol 146:451–464

Spatz WB, Tigges J, Tigges M (1970) Subcortical projections, cortical associations, and some intrinsic interlaminar connections of the striate cortex in the squirrel monkey (Saimiri). J Comp Neurol 140:155–174

Spoerri PE, Spoerri O, Glees P (1979) Reacting ultrastructure of the human oligodendrocyte. A study of cerebral cortex distant to brain tumours. Acta Neurochir (Wien) 46:45–52

Szentagothai J (1971) Some geometrical aspects of the neocortical neuropil. Acta Biol Acad Sci Hung 22:107–124

Tigges J, Tigges M, Kalaha CS (1973) Efferent connections of area 17 in Galago. Am J Phys Anthropol 38:393–397

Tigges J, Spatz WB, Tigges M (1974) Efferent cortico-cortical fiber connections of area 18 in the squirrel monkey (Saimiri). J Comp Neurol 158:219–236

Tigges J, Tigges M, Perachio AA (1977) Complementary laminar terminations of afferents to area 17 originating in area 18 and in the lateral geniculate nucleus in squirrel monkey. J Comp Neurol 176:87–100

Tigges M, Tigges J (1979) Types of degenerating geniculo-cortical axon terminals and their contribution to layer IV of area 17 in the squirrel monkey (Saimiri). Cell Tissue Res 196:471–486

Tigges M, Bos J, Tigges J, Bridges E (1977) Ultrastructural characteristics of layer IV neuropil in area 17 of monkeys. Cell Tissue Res 182:39–59

Tömböl T (1974) An electron microscopic study of the neurons of the visual cortex. J Neurocytol 3:525–531

Tömböl T (1978) Some Golgi data on visual cortex of the rhesus monkey. Acta Morphol Acad Sci Hung 26:115–138

Tömböl T, Hadju F, Somogyi G (1975) Identification of the Golgi picture of the layer VI cortico-geniculate projection neurons. Exp Brain Res 24:107–110

Trojanowski JQ, Jacobson S (1977) The morphology and laminar distribution of cortico-pulvinar neurons in the rhesus monkey. Exp Brain Res 28:51–62

Tusques J, George Y, Couderc M, Roch M (1973) Ultrastructure de l'oligodendroglie de l'ecore cerebrale humaine. Bull Assoc Anat (Nancy) 57:939–948

Valverde F (1971) Short axon neuronal subsystems in the visual cortex of the monkey. Int J Neurosci 1:181–197

Valverde F (1976) Aspects of cortical organization related to the geometry of neurons with intracortical axons. J Neurocytol 5:509–529

Valverde F (1977) Lamination of the striate cortex. J Neurocytol 6:483–484

Van der Loos H (1965) The improperly orientated pyramidal cell in the cerebral cortex and its possible bearing on problems of growth and cell orientation. Bull Johns Hopkins Hosp 117:228–250

Varon SS, Somjen GG (1979) Neuron-glia interaction. Neurosci Res Program Bull 17:6–239

Vaughan DW, Peters A (1973) A three dimensional study of layer I of the rat parietal cortex. J Comp Neurol 149:355–370

Vaughan DW, Peters A (1974) Neuroglial cells in the cerebral cortex of rats from young adulthood to old age: an electron microscope study. J Neurocytol 3:405–429

Vogt C, Vogt O (1919) Allgemeinere Ergebnisse unserer Hirnforschung. J Psychol Neurol (Leipzig) 25:279–462

Von Bonin G (1942) The striate area of primates. J Comp Neurol 77:405–429

Von Bonin G, Mehler WR (1971) On columnar arrangement of nerve cells in cerebral cortex. Brain Res 27:1–9

Von Economo C, Koskinas GN (1925) Die Cytoarchitektonik der Hirnrinde des erwachsenen Menschen, Springer, Vienna

Winkelmann E, Brauer K, Berger U (1975) Zur columnaren Organisation von Pyramidenzellen im visuellen Cortex der Albinoratte. Z Mikrosk Anat Forsch 89:239–256

White FP (1980) Protein synthesis in rat telencephalon slices: high amounts of newly synthesized protein found in association with brain capillaries. Neuroscience 5:173–178

Winfield DA, Powell TPS (1976) The termination of thalamocortical fibres in the visual cortex of the cat. J Neurocytol 5:269–281

Wolff J (1968) Die Astroglia im Gewebsverband des Gehirns. Acta Neuropathol (Berl) [Suppl] IV: 33–39

Wong-Riley M (1978) Reciprocal connections between striate and prestriate cortex in squirrel monkey as demonstrated by combined peroxidase histochemistry and autoradiography. Brain Res 147:159–164

Wong-Riley M (1979) Columnar cortico-cortical interconnections within the visual system of the squirrel and macaque monkeys. Brain Res 162:201–217

Woolsey CN, Akert K, Benjamin RM, Leibowitz H, Welker MJ (1955) Visual cortex of the marmoset Callithrix jacchus. Fed Proc 14:106

Wree A, Zilles K, Schleicher A (1980) Analyse der laminären Struktur der Area striata mit verschiedenen stereologischen Methoden. Verh Anat Ges 74:727–728

Zeki SM (1971) Convergent input from the striate cortex (area 17) to the cortex of the superior temporal sulcus in the rhesus monkey. Brain Res 28:338–340

Zeki SM (1976) The projections to the superior temporal sulcus from area 17 and 18 in the rhesus monkey. Proc R Soc Lond [Biol] 193:199–207

Subject Index

Advances in Anatomy Embryology and Cell Biology

Editors: F. Beck, W. Hild, J. van Limborgh,
R. Ortmann, J. E. Pauly, T. H. Schiebler

A Selection

Volume 68
A. A. M. Gribnau, L. G. M. Geijsberts
Developmental Stages in the Rhesus Monkey (Macaca mulatta)
1981. 27 figures. VI, 84 pages.
ISBN 3-540-10469-0

Volume 69
L. Záborszky
Afferent Connections of the Medial Basal Hypothalamus
1982. 31 figures. VIII, 107 pages.
ISBN 3-540-11076-3

Volume 70
W. Pfaller
Structure Function Correlation on Rat Kidney
Quantitative Correlation of Structure and Function in the Normal and Injured Rat Kidney
1982. 23 figures. VIII, 106 pages.
ISBN 3-540-11074-7

Volume 71
L. Thuneberg
Interstitial Cells of Cajal: Intestinal Pacemaker Cells?
1982. 94 figures. Approx. 120 pages.
ISBN 3-540-11261-8

Volume 72
H. Breuker
Seasonal Spermatogenesis in the Mute Swan (Cygnus olor)
1982. 30 figures. VII, 94 pages.
ISBN 3-540-11326-6

Volume 73
G. Zweers
The Feeding System of the Pigeon (Columba livia L.)
1982. 45 figures. VII, 108 Seiten
ISBN 3-540-11332-0

Volume 74
J. Altman, S. A. Bayer
Development of the Cranial Nerve Ganglia and Related Nuclei in the Rat
1982. 64 figures. VII, 90 Seiten
ISBN 3-540-11337-1

Volume 75
V. Grouls, B. Helpap
The Development of the Red Pulp in the Spleen
1982. 37 figures.
Approx. 70 pages.
ISBN 3-540-11408-4

Volume 76
P. Kugler
On Angiotensin-Degrading Aminopeptidases in the Rat Kidney
Translated from the German by T. Telger
1982. 88 figures. Approx. 112 pages.
ISBN 3-540-11452-1

Springer-Verlag
Berlin
Heidelberg
New York

Techniques in Neuroanatomical Research

Editors: C. Heym, W.-G. Forssmann

1981. 165 figures. XIII, 395 pages. ISBN 3-540-10686-3

Contents: General Research Methods in Neuroanatomy. – Light Microscopical Research Methods in Neuroanatomy. – Electron Microscopical Research Methods in Neuroanatomy. – Investigation of Living Nervous Tissue. – Subject Index.

Techniques in Neuroanatomical Research is a detailed presentation of selected methods the neurobiologist needs for conducting morphological studies. The formulae and practical introductions provided for each technique allow even non-morphologists to acquire them easily. Each chapter concludes with an extensive bibliography to enable the reader to broaden his knowledge while at the same time introducing him to the works of leading scientists in the field.

The Renin Angiotensin System in the Brain

A Model for the Synthesis of Peptides in the Brain

Editors: D. Ganten, M. Printz, M. I. Phillips, B. A. Schölkens

1982. 108 figures, 46 tables. XVII, 385 pages. ISBN 3-540-11344-4

Contents: Renin and Converting Enzyme. – Angiotensinogen, Angiotensin, Angiotensin Receptors. – Central Effects of Angiotensin. – Nomenclature of the Renin – Angiotensin System. – Nomenclature of Experimental Hypertension. – Subject Index. – Author Index.

The existence of an endogenous brain renin angiotensin system has been a subject of controversy and intensive research over the past ten years. Definite proof of the existence of the component of this enzyme peptide system is provided for the first time in this monograph. Leading researchers present the state of the art on the brain renin angiotensin system based on biochemical, pharmacological, physiological and endocrinological studies. The significance of these studies lies in their contribution to our understanding of the synthesis, distribution, physiology, and function of neuropeptides. Divergent functions can now be attributed to the brain renin angiotensin system – ranging from its pathophysiologic role in hypertension to its modification on thirst and memory – and making the information contained in this volume of profound interest to both basic scientists as well as to clinicians in internal medicine and psychiatry.

Springer-Verlag
Berlin
Heidelberg
New York